MODELLING, APPLICATIONS AND APPLIED PROBLEM SOLVING
Teaching Mathematics in a Real Context

MATHEMATICS AND ITS APPLICATIONS

Series Editor: G. M. BELL, Professor of Mathematics,
King's College London (KQC), University of London

STATISTICS, OPERATIONAL RESEARCH AND COMPUTATIONAL MATHEMATICS

Editor: B. W. CONOLLY, Emeritus Professor of Mathematics (Operational Research), Queen Mary College, University of London

Mathematics and its applications are now awe-inspiring in their scope, variety and depth. Not only is there rapid growth in pure mathematics and its applications to the traditional fields of the physical sciences, engineering and statistics, but new fields of application are emerging in biology, ecology and social organization. The user of mathematics must assimilate subtle new techniques and also learn to handle the great power of the computer efficiently and economically.

The need for clear, concise and authoritative texts is thus greater than ever and our series will endeavour to supply this need. It aims to be comprehensive and yet flexible. Works surveying recent research will introduce new areas and up-to-date mathematical methods. Undergraduate texts on established topics will stimulate student interest by including applications relevant at the present day. The series will also include selected volumes of lecture notes which will enable certain important topics to be presented earlier than would otherwise be possible.

In all these ways it is hoped to render a valuable service to those who learn, teach, develop and use mathematics.

Mathematics and its Applications

Series Editor: G. M. BELL, Professor of Mathematics, King's College London (KQC), University of London

Series continued at back of book

MODELLING, APPLICATIONS AND APPLIED PROBLEM SOLVING
Teaching Mathematics in a Real Context

WERNER BLUM
Professor of Mathematics and Mathematics Education
University of Kassel, FRG

M. NISS
Professor of Mathematics and Mathematics Education
University of Roskilde, Denmark

I. HUNTLEY
Head of Department of Mathematics
Sheffield City Polytechnic, UK

ELLIS HORWOOD LIMITED
Publishers · Chichester

Halsted Press: a division of
JOHN WILEY & SONS
New York · Chichester · Brisbane · Toronto

First published in 1989 by
ELLIS HORWOOD LIMITED
Market Cross House, Cooper Street,
Chichester, West Sussex, PO19 1EB, England
The publisher's colophon is reproduced from James Gillison's drawing of the ancient Market Cross, Chichester.

Distributors:

Australia and New Zealand:
JACARANDA WILEY LIMITED
GPO Box 859, Brisbane, Queensland 4001, Australia

Canada:
JOHN WILEY & SONS CANADA LIMITED
22 Worcester Road, Rexdale, Ontario, Canada

Europe and Africa:
JOHN WILEY & SONS LIMITED
Baffins Lane, Chichester, West Sussex, England

North and South America and the rest of the world:
Halsted Press: a division of
JOHN WILEY & SONS
605 Third Avenue, New York, NY 10158, USA

South-East Asia
JOHN WILEY & SONS (SEA) PTE LIMITED
37 Jalan Pemimpin # 05–04
Block B, Union Industrial Building, Singapore 2057

Indian Subcontinent
WILEY EASTERN LIMITED
4835/24 Ansari Road
Daryaganj, New Delhi 110002, India

© **1989 W. Blum, M. Niss and I. Huntley/Ellis Horwood Limited**

British Library Cataloguing in Publication Data
Blum, W. (Werner), *1945–*
Modelling, applications and applied problem solving.
1. Education. Curriculum subjects: Mathematics. Teaching
I. Title   II. Niss, M.   III. Huntley, I.   IV. Series
510'.7

Library of Congress Card No. 89–19889

ISBN 0–7458–0633–3 (Ellis Horwood Limited)
ISBN 0–470–21570–4 (Halsted Press)

Printed in Great Britain by Hartnolls, Bodmin

Table of contents

Section E. *National Surveys*

Preface

This volume contains selected papers contributed to the Theme Groups on Problem Solving, Modelling and Applications (Group T3; Chief Organiser: Mogens Niss) and on Mathematics and Other Subjects (Group T6; Chief Organiser: Werner Blum) at the Sixth International Congress on Mathematics Education (ICME-6), which was held in Budapest, Hungary, July/August, 1988.

Problem solving and the relations between mathematics and the rest of the world – which are all brought about through *applications, modelling* and *applied problem solving* – as well as the parts they play in mathematics teaching from school to university, have attracted increasing attention during the history of the ICMEs (ICME-1, Lyon 1969; -2, Exeter 1972; -3, Karlsruhe 1976; -4, Berkeley 1980; -5, Adelaide 1984; see the various congress proceedings). This is also the case with other international fora in mathematics education, especially the series of International Conferences on the Teaching of Mathematical Modelling and Applications, the ICTMAs (see the conference proceedings, edited by Berry et al and Blum et al, and published in the same series as the present volume).

In Budapest, the complex of topics forming the theme of the present book was divided into two parts. Group T3 dealt with the *processes* of pure and applied mathematical *problem solving* and with the links between mathematics and the world outside established by *modelling* and *applications*, whereas Group T6 was focussed on the *relations* between mathematics and the real world, particularly *other* school or university *subjects*, with special reference to mathematics as a service subject. (Details may be found in the two Theme Group reports written by the Chief Organisers in the Proceedings of ICME 6 edited by Hirst and Hirst, János Bolyai Mathematical Society, Budapest 1988).

Parts of the theoretical basis for the work in the two Theme Groups were laid by a joint *survey* lecture, given at ICME-6 by the two Chief Organisers. The first article in this book contains a shortened version of this lecture. In the article, certain basic concepts are defined, and the most important arguments for and against giving problem solving, modelling and applications a strong position in mathematics instruction are outlined. The main portion of the paper is spent in identifying and reviewing the present state, recent trends and prospective lines of development in the field, and by discussing consequences for methods, curricula and assessment in mathematics instruction.

The following articles contain reports on concrete *examples*, *materials* or *projects*, and on *empirical investigations*, as well as *theoretical reflections* on issues related to modelling, applications and applied problem solving in mathematics teaching at school and university. The reader will note that as far as problem solving is concerned, its *applied* aspects are in focus.

The articles are grouped into *five sections*. This grouping is not intended as a strict and clear-cut classification of the contributions, but rather as a pragmatic ordering which should aid the reader in looking for particular items of interest.

The five contributions of **Section A** concentrate on *foundational* and *theoretical aspects* of the field, in particular as regards methodological, philosophical and historical issues.

Section B contains *examples* and *materials* for modelling, applications and applied problem solving in mathematics instruction at the *primary* (two articles) and at the *lower secondary* school level (six articles). **Section C** contains the same sort of contributions, related to mathematics instruction at the *upper secondary* (three articles) and the *tertiary* level (three articles).

The six articles in **Section D** all report on *empirical investigations* into modelling, applications and applied problem solving as objects of mathematics instruction. All educational levels from lower secondary to tertiary are covered by the research presented in this section.

The final **Section E** contains four *national surveys* (from Australia, Denmark, Finland and the Federal Republic of Germany) on the state of modelling, applications and applied problem solving in the secondary school mathematics curricula of the respective countries.

We would like to finish this preface by extending our thanks to all the participants of Theme Groups T3 and T6 at ICME-6 who have contributed, in one way or another, to the realisation of this volume. We hope that the book will help further to advance discussions, research and development concerning modelling, applications and applied problem solving in the learning and teaching of mathematics. Last but not least we wish to thank the staff at Sheffield City Polytechnic, in particular Beverley Lougher, Maggie Bedingham and Richard Gibson, for their patient and tedious work with preparing the camera-ready manuscript and with correcting the English in the many places where it was necessary.

Werner Blum, Mogens Niss, Ian Huntley

CHAPTER 1

Mathematical Problem Solving, Modelling, Applications, and Links to Other Subjects – State, Trends and Issues in Mathematics Instruction[1]

W. Blum
Kassel University, FR Germany
M. Niss
Roskilde University, Denmark

ABSTRACT
The paper will consist of three parts. In **part one** we shall present some background considerations which are necessary as a basis for what follows. We shall try to clarify some basic concepts and notions, and we shall collect the most important arguments (and related goals) in favour of problem solving, modelling and applications to other subjects in mathematics instruction. In the main **part two** we shall review the present state, recent trends, and prospective lines of development, both in empirical or theoretical research and in the practice of mathematics instruction and mathematics education, concerning problem solving, modelling, applications and relations to other subjects. In particular, we shall identify and discuss four major trends: a widened spectrum of arguments, an increased globality, an increased unification, and an extended use of computers. In the final **part three** we shall comment upon some important issues and problems related to our topic.

1. BACKGROUND CONSIDERATIONS

1.1 Clarification of basic concepts and notions
We shall commence our paper by clarifying some basic concepts and notions such as *problem*, *modelling* or *application*. By no means are we

pretending to present an exhaustive epistemological treatment of these concepts. Rather, this section presents a pragmatic attempt to give some working definitions.

By a *problem* we mean a situation which carries with it certain open questions that challenge somebody intellectually who is not in immediate possession of direct methods sufficient to answer the questions. As to *mathematical* problems, there are two kinds. It is characteristic of an *applied problem* that the situation and the questions defining it belong to some segment of the real world and allow some mathematical concepts, methods and results to become involved. By *real world* we mean the rest of the world outside mathematics, ie school or university subjects or disciplines different from mathematics, or everyday life and the world around us. In contrast, with a *purely mathematical problem* the defining situation is entirely embedded in some mathematical universe.

Now, *problem solving* simply refers to the entire *process* of dealing with a problem – pure or applied – in attempting to solve it.

In *mathematics education* problem solving is considered in two ways. (i) As an object of *research* on issues such as How is problem solving related to other aspects of thinking mathematically? (ii) In relation to *mathematics instruction*, where issues concerning the inclusion and implementation of problem solving in mathematics curricula are addressed. In this paper we have to confine ourselves to treating only the *second* aspect.

Next, we are going to look at the *applied problem solving process* in more detail. The following way of describing the interplay between the real world and mathematics is well known and is by no means our invention (see for example Blechman et al 1984, Steiner 1976, Pollak 1979 or Blum 1985). The starting point is an applied problem or, as we shall also say, a *real problem situation*. This situation has to be simplified, idealised, structured and to be made more precise by the problem solver according to his/her interests. This leads to a *real model* of the original situation.

The real model has to be *mathematised*, ie its data, concepts, relations, conditions and assumptions are to be translated into mathematics. Thus, a *mathematical model* of the original situation results. While *mathematisation* is the process from the real model into mathematics, we mean by *modelling* or *model building* the entire process leading from the original real problem situation to a mathematical model.

It has proved appropriate to distinguish between different kinds of models. If, for example, economic items such as interest or taxes are considered mathematics particularly serves to establish certain norms involving value judgements. Here it is a matter of *normative* models. If physical phenomena such as planetary motions or radioactive decay are considered, however, mathematics serves primarily to describe and explain the respective situation. Here it is a matter of *descriptive* models.

The applied problem solving process continues by work *within* mathematics through which certain *mathematical results* are obtained. These results have to be retranslated into the real world, ie to be *interpreted* in relation to the original situation. In doing so the problem solver also *validates* the model, ie decides whether it is justified to use it for the purposes for which it was built. When validating the model, *discrepancies* of various kinds may occur which lead to a modification of the model or to its replacement by a new one. In other words, the problem solving process may require going round the loop *several* times.

Besides such complex problem solving processes – which are rare in mathematics instruction – there are abbreviated and restricted links between mathematics and reality which are much more frequently found. On the one hand a *direct application* of already developed standard mathematical models to real situations with a mathematical content, on the other hand a dressing up of purely mathematical problems in the words of another discipline or of everyday life. Such word problems often give a distorted picture of reality. This is sometimes done deliberately in order to serve instructional purposes.

It is common practice to use the terms *application* of mathematics to denote all the above-mentioned ways of bringing the real world into a relationship with mathematics. In this sense real problem situations can also be called *applications*. Eventually, mathematical models or, more generally, every piece of mathematics which in some way is or may be related to the real world, can be seen as belonging to *applied mathematics*.

We will conclude this first section by some remarks on which kinds of *mathematics instruction, in relation to other subjects*, we shall consider in this paper. When speaking about mathematics instruction we regard it as taking place within a given segment of an existing educational system. Here, firstly, mathematics instruction may essentially serve two different *purposes*:

(a) to provide students with knowledge and abilities concerning *mathematics* as a subject in itself,

(b) to provide students with knowledge and abilities concerning *other subjects*, to which mathematics is supposed to have actual or potential services to offer.

Secondly, the *organisational framework* of mathematics instruction may take two different shapes.

(1) Mathematics may be taught as a *separate subject*.

(2) Mathematics may be taught as a part of and *integrated* in other subjects.

Thirdly, we distinguish between mathematics instruction in different *educational histories*.

(1) Mathematics in *schools* offering *general* education.

(2) Mathematics in *vocational* education.

(3) Mathematics in *university* courses for future *mathematicians* or mathematics teachers.

(4) Mathematics as a *service* subject in *university* courses for future scientists, engineers, economists and so on.

Now we can illustrate the situation by a matrix.

purpose / organisation	(a) focus on **mathematics**	(b) focus on **other subjects**
(1) mathematics as a **separate subject**	(1a) examples: ①,③	(1b) examples: ④,②, partly ①
(2) mathematics **integrated** in other subjects	(2a) examples:	(2b) examples: ②, partly ④

integrated curricula

In all the cells of this matrix, relations between mathematics and other subjects may have a part to play. Even if, as in (1a), the purpose of mathematics instruction is to elucidate mathematics as a subject, it may be highly relevant to incorporate mathematical applications and modelling; we shall present arguments for that in the next section 1.2. Possible relations shown in the matrix also include truly *integrated* curricula (second row of the matrix), both in school and in university, whereby teaching and learning is taking place in an interdisciplinary way. There are, however, only a few experiences of using this approach.

When we talk here about *mathematics instruction* we always have the *whole* matrix in mind, providing that it is possible to distinguish segments of instruction with mathematics as an explicit object of attention. We shall only abstain from regarding instances of (2b), where mathematics is totally integrated in other subjects.

1.2 Review of arguments

Throughout the history of mathematics education, the inclusion of aspects of applications and - more recently - problem solving and modelling in mathematics instruction has been regularly advocated by various individuals and quarters, and in periods also realised in some curricula. A review of representative literature on mathematics education shows that basically **five arguments** for such an inclusion have been invoked over the years.

1. **The formative argument** emphasises the application of mathematics and the performing of mathematical modelling and problem solving as suitable means for developing *general competences and attitudes* with students, in particular orientated towards fostering overall explorative, creative and problem solving capacities, as well as open-mindedness and self-reliance.

2. **The critical competence argument** focusses on preparing students to live and act with integrity as private and social citizens, possessing a *critical competence* in a society which is being increasingly influenced by the utilisation of mathematics through applications and modelling. The aim of such a competence is to enable students to see and judge independently, to recognise and understand representative examples of actual uses of mathematics.

3. **The utility argument** emphasises that mathematics instruction should prepare students to *utilise mathematics* for solving problems in or describing aspects of specific extra-mathematical areas and situations, whether referring to other subjects or occupational contexts (mathematics as a service subject) or to the actual or future everyday lives of students. In other words, mathematics instruction should enable students to *practice applications, modelling and problem solving* in a variety of contexts.

4. **The picture of mathematics argument** insists that it is an important task of mathematics education to establish with students a *rich and comprehensive picture of mathematics* in all its facets, as a science, as a field of activity in society and culture. Since applications, modelling and problem solving constitute an essential component in such a picture, this component should be allotted an appropriate position in mathematics curricula.

5. **The prompting mathematics learning argument** emphasises that the incorporation of problem solving, applications and modelling aspects and activities in mathematics instruction is well suited to assist students in acquiring, learning and keeping mathematical concepts, notions, methods and results, by providing motivation for and relevance of mathematical studies. Such work also contributes to exercise students in thinking mathematically, and in selecting and performing mathematical techniques within and outside mathematics.

Along with the arguments put forward *in favour* of including applications, modelling and problem solving in mathematics instruction, also *counter-arguments* exist.   Some of them will be discussed in section 3.1 of the present paper.   So, this is not the occasion to offer a detailed discussion of the arguments *for* versus the arguments *against* assigning applications, modelling and problem solving significant roles in mathematics instruction.   Not surprisingly, to us, the collection of arguments for outweigh the counter-arguments.   Let us sum up our reasoning in two points.   (I)   Considering the fact that substantial mathematics education is no longer reserved for a small minority of people, but is now being given to an ever larger proportion of the population, mathematics instruction at all levels has to deal also with the role and use of mathematics in the world outside the realm of mathematics itself.   (II)   Mastering mathematics can no longer be considered equivalent to knowing a set of mathematical *facts*.   It requires also the mastering of mathematical *processes*, of which problem solving – in its broadest understanding – occupies a predominant position.

This, however, should not be taken to imply that proper mathematical knowledge, proficiency and insight become less important. On the contrary – the more widely and extensively mathematics is being activated and used, the more necessary genuine mathematical knowledge becomes for the understanding, evaluation and judging of its use.

If we agree that problem solving, modelling and applications to other disciplines should be granted important positions in mathematics instruction, then their inclusion may pursue one or more of several **overall goals**, among which the following three seem to be particularly significant (cf Niss 1989).

Goal 1: Students should be able to *perform* applicational/modelling/ problem solving processes.

Goal 2: Students should *acquire knowledge* of existing models and applications of mathematics, and/or of characteristic aspects of applications/modelling/problem solving processes.

Goal 3: Students should be able to *analyse and assess* critically, given examples of applications, models, modelling and problem solving.

2.   PRESENT STATE AND CURRENT TRENDS

Problem solving, modelling, relations with and applications to other disciplines are no longer upstarts on the stage of mathematics education, fighting for attention and recognition.   Aspects of these topics were on the agendas of ICMEs–3 (see Pollak 1979), –4 (see Bell 1983) and –5 (cf Lesh et al 1986), and now on that of ICME–6 (cf Niss 1988 and Blum 1988), as major themes of increasing significance to mathematics education and instruction, and of increasing interest to the mathematics education community in the world. This interest has manifested itself in other forums as well, in several conferences - especially in the series of biennial International Conferences on the Teaching of Mathematical Modelling and Applications (ITCMA) held in 1983, 1985, 1987 and 1989 (see Berry et al 1984, 1986, 1987 and Blum et al 1989) - and in a host of publications (see section 3.2). In the following, we shall concentrate on reviewing the present state and current trends, with an emphasis on the last half a decade.

There are two major aspects of studies and activities concerning problem solving, modelling, applications, and the relations between mathematics and other subjects: an aspect of empirical and theoretical *research*, and an aspect of *practice* defined as the actual implementation of mathematics education in the educational system. In the term research we include not only the creation of new positive (often empirical) knowledge, but also *systematic reflections* - whether mathematical, philosophical, psychological, sociological or whatever - on mathematics education, including the development of new *types* of curricula based on such reflections. Also the term practice should be taken in a wide sense, referring not only to everyday classroom teaching practice. It comprises in our definition all elements in the actual implementation of mathematics education including, for instance, the devising and carrying out of *specific* entire curricula, or curriculum components, the writing of textbooks, creation of teaching materials and so on.

In the present brief survey we shall now concentrate on outlining **four major trends**.

2.1 Trend 1 : A widened spectrum of arguments

It is probable that all five arguments mentioned in section 1.2 for incorporating problem solving, modelling and applications in other areas in mathematics instruction have been invoked in some form or another during the last century and a half. Yet, traditionally, the predominant arguments have been only two of the five: the utility argument (3) and the promoting mathematics learning argument (5). From the late sixties the formative argument (1) began to be called upon frequently as well. (In Germany and France this happened earlier.) Eventually, during the last decade, the picture of mathematics argument (4) and, more recently, the critical competence argument (2) have gained momentum too.

So presently *all five arguments* are seen to be put forward to motivate that mathematics instruction at all levels should deal with problem solving, modelling, applications, and relations to other subjects. This widening of the spectrum of arguments is significant in that it places the issue of problem solving, modelling and applications where it belongs: not merely amongst tactical devices to improve the situation for *traditional* mathematics instruction, but as an *integral part* of the discussion of mathematics education as a whole.

2.2   Trend 2 : Increasing globality

Over the last handful of years, an increasing globality of the theoretical and practical interest in and activities of problem solving, modelling and applications to other areas and subjects can be detected. This is true both if globality is taken in a *geographical* sense, and if it refers to *internal* aspects of mathematics instruction.

(i)   In recent years, all over the world, a growing number of mathematics programmes at different levels are including problem solving, applications, models and modelling in extra-mathematical areas and subjects.   Similarly, an increasing amount of cooperation between mathematics and other subjects can be observed.   We may liken the development with a logistic growth process.   The process is (still) in the steep segment of the curve but there are indications that it is approaching the segments of slower growth corresponding to gradual saturation.

(ii)   At ICME-5 it was stated (cf Niss 1987) that the distance between the *front* of research, development and practice in applications and modelling, on the one hand, and applications and modelling in the *mainstream* of mathematics instructions on the other hand, was very large;   even more so with problem solving, at least as regards post-elementary mathematics instruction (ie instruction dealing with mathematics beyond arithmetic and simple computational geometry). Although the distance between the forefront and the mainstream is still very *large* it is *being reduced*.   An indication of this may be found in the increasing inclusion of applications and model cases in textbooks from school to university.

(iii)   Concurrently with the dissemination of problem solving, applications, modelling and increased cooperation between mathematics and other subjects into the mainstream of mathematics instruction, the *rate of innovation at the front* seems to be *decreasing*.   Thus, with minor exceptions, neither the spectrum of application areas nor the spectrum of mathematical topics dealt with in mathematics instruction have become fundamentally extended since ICME-5.   Similarly, the paradigms of research on problem solving, applications and modelling have not really changed during the last half decade.   This is hardly surprising.   For if it is true, as we claim, that what has been happening in the last half decade or so is that ideas and materials developed at the front are now spreading rather smoothly into mainstream mathematics instruction, without having been confronted, as yet, with an extensive and massive amount of experiences indicating problems and calling for revision, there is no great demand and incentive for front-line innovation at the moment.

2.3   Trend 3 : Increasing unification

Earlier, the conceptual and other relations between problem solving, modelling, models and applications, and cooperation between mathematics and other subjects, were seldom made clear (and this is still the case to some degree). Nevertheless, it is possible to distinguish over the years separate tendencies and groupings in mathematics education, concentrating – respectively – on applications, modelling, problem solving, mathematics as a service subject, and cooperation between mathematics and other subjects.

In recent years these different tendencies and groupings have become more amalgamated. We may speak of an *increasing unification* of the field. This should not be taken to imply that the different components of the field have merged into a homogeneous whole. What has happened is that all quarters have widened their horizons and have become increasingly aware of and interested in areas of contact and interaction with the other quarters. The development of the field could be described as a transition from a discrete towards a continuous spectrum of interest and activity.

(i) Although the trend of increasing unification in the field of problem solving, modelling, applications, and cooperation between mathematics and other subjects is pretty international in character, national/regional differences in focus and emphasis exist. So, for example, the *psychological* aspects of *problem solving* are widely investigated in empirical and (to a lesser extent) theoretical studies conducted in the USA; also the USSR, Canada and Israel have strong traditions in this area. The *pragmatic* aspect of problem solving, including its implementation in mathematics curricula – mainly primary and lower secondary ones – are receiving considerable attention in the UK and Australia, in particular as regards *applied* problem solving and so called *investigations* in less structured situations, and in Finland.

The majority of contributions concerning the *theoretical* and *philosophical* aspects of *applications and modelling* in an educational context come from the FRG, Austria, the Netherlands, Denmark, France and India (for a discussion see Kaiser–Messmer 1986, vol. I.) For a long time the leading country as regards the *pragmatic* aspects of applications and modelling has been the UK, but very interesting work is being done in many other countries too, for example the USA, the Netherlands, Australia, Italy and FRG. Countries in which a fair amount of valuable pragmatic cooperation between mathematics and other subjects, primarily in upper secondary school, has been carried out, are Denmark and the Netherlands.

(ii) Relations between *mathematics and physics* as sciences have always been very intimate and profound, so much so that many mathematical concepts were established in close connection with attempts at describing and understanding physical systems. This intimate relationship used to be reflected in mathematics instruction at most levels. However, it seems that over the last decade the *relations have become weakened* in many places. This is due to the *opening* of mathematics instruction to new applicational areas. There are several reasons for this opening, among others the following.

- A growing proportion of students enter professions utilising mathematics in non-physical contexts.

- As post-elementary school mathematics education is being given to an increasing number of children and youngsters to whom it is their final formal encounter with mathematics, the range of areas to which they see mathematics being applied has to encompass more than just physics in order to be relevant to their interests and their lives in society.

- Since mathematical applications and modelling in areas outside physics generally rely on less involved and demanding extra-mathematical theory than is the case with physics, such areas often provide opportunities to activate mathematics in ways that are more easily accessible to the majority of students.

In our view, this opening is both necessary and desirable. However, at the school level it is very important that a *close contact between mathematics and physics* be *maintained*, although unavoidably this has to be on a smaller scale than was usual in earlier days. It may be said, somewhat paradoxically perhaps, that the more mathematics is being applied to areas and subjects outside physics, the more essential it is to have access to representative cases from physics to shed light on possibilities, conditions, difficulties and pitfalls of applications and modelling in fields with smaller degrees of well-established mathematisation.

2.4 Trend 4 : An extended involvement of computers

For several years it has been evident that computers form a highly powerful tool for the numerical and graphical treatment of mathematical applications and models. Not only do computers allow for greater detail, accuracy, and rapidity in calculations, for the handling of more data, for examining the effects of changing terms or parameters, for providing better illustrations, and so on, but in many cases they are simply indispensable for a given mathematical model to be accessible or realistic. As soon as technically feasible and sufficiently inexpensive microcomputers become available on a large scale, they entered mathematics instruction too and were utilised, as they still are, to treat mathematical models and applications. Thus there is a trend to *quantitatively extend* the involvement of computers in mathematics instruction to deal with applications and models, a trend which is hardly surprising to anyone.

Recently, however, a somewhat different trend of *qualitative extension* of the involvement of computers in problem solving, modelling and applications has manifested itself. Quite a fair amount of software has been developed to offer various sorts of assistance to the *process* of problem solving or modelling, or to the process of applying mathematics to various areas. Some kinds of software offer opportunities to explore certain types of problem solving or modelling situations, other kinds provide interactive tools for *building* models or for *investigating* model behaviour within certain standard model universes (eg differential equations). It is a significant characteristic of this new way of implicating computers in problem solving, modelling and applications, that they may be utilised without knowing and understanding the mathematics involved.

This development may have several effects which we shall comment on in section 3.3.

3. ISSUES AND PROBLEMS
In this final part of our paper we shall identify and comment on some of the most important issues and problems in relation to problem solving, modelling, applications, and the interplay between mathematics and other subjects. We have to confine ourselves here to outlining briefly a few main points. Emphasis will be laid on *curricular* and *instructional* aspects, whereas less attention will be paid to research aspects in a strict sense.

3.1 Obstacles to problem solving, modelling and applications
In spite of all the good arguments in favour of problem solving, modelling, applications and links to other subjects in mathematics teaching, collected in section 1.2, these items often still do not play as important a role in mainstream mathematics instruction at school and university as we would wish (see for example Burkhardt 1983). This is not due to ill−will or incompetence of teachers but to certain objective *obstacles*. Such obstacles have been well known to mathematics educators for a long time (see for example Pollak 1979, Blum 1985 or Niss 1987), but they still exist. We shall refer to three kinds.

(A) Obstacles from the point of view of instruction
Many mathematics teachers from school to university are afraid of not having enough *time* to deal with problem solving, modelling and applications in addition to the wealth of compulsory mathematics included in the curriculum. Furthermore, some teachers even doubt whether applications and connections to other subjects belong to mathematics instruction at all, because such components tend to *distort* the aesthetic purity, the beauty and the context−free universality of mathematics.

(B) Obstacles from the learner's point of view
Problem solving, modelling and applications to other disciplines make the mathematics lessons unquestionably more *demanding* and less predictable for learners. Routine mathematical tasks such as calculations are more popular with many students, because they are much easier to grasp and can often be solved merely by following certain recipes.

(C)   Obstacles from the teacher's point of view

Problem solving and references to the world outside mathematics make instruction more *open* and more *demanding* for teachers and make it more *difficult* to assess students' achievements.   Moreover, many teachers do not *feel able* to deal with applied examples which are not taken from the subjects they have studied themselves.   Very often teachers simply either do not *know* enough examples or they do not have enough *time* to adapt examples to the actual class.

Nevertheless, in the light of the arguments put forward in favour of problem solving, modelling, applications and connections to other subjects we should continue to make every effort to *overcome* these obstacles. This could be done both by adequate pre-service and in-service teacher education, to equip teachers with knowledge, abilities, experiences, and in particular with attitudes to cope with the demands of teaching problem solving, modelling and applications, and by stimulating every kind of contact and cooperation between mathematics teachers at school and university and their colleagues in other subjects.   We should urge that *problem solving and relations to the real world become and remain essential parts* of mathematics instruction at all levels, even in spite of all the difficulties mentioned.

A few more remarks on the third obstacle.   Are there, for each level of mathematics instruction from school to university, really *enough cases* of problem solving, modelling and applications to other subject areas, suitable for teaching?   For the elementary school level, the answer is obviously Yes.   For the secondary and the tertiary level, our answer here is Yes again (perhaps in contrast to the answer we would have given 20 years ago), and we shall proceed by mentioning a few important materials and resources.

3.2 Materials and resources

Firstly, we mention the various *references* to materials and literature given in Pollak (1979) and in Bell (1983).   We also refer to the extensive *bibliography* of Kaiser et al 1982 (with a supplement to appear in 1990).   From among the various current *curriculum projects* we shall list only a few.

(1)   From the USA

- The High School Mathematics and its Applications Project (HIMAP), and the Undergraduate Mathematics and its Applications Project (UMAP), both coordinated and published by the Consortium for Mathematics and its Applications (COMAP), directed by Solomon Garfunkel and Laurie Aragon.

- The University of Chicago School Mathematics Project (UCSMP), directed by Zalman Usiskin (see Usiskin 1989).

- The Committee on Enrichment Modules and its continuation Contemporary Applied Mathematics, directed by Clifford Sloyer and others at the University of Delaware (see Sloyer 1989).

(2) From Great Britain

- Several projects at the Shell Centre for Mathematical Education, University of Nottingham, directed by Hugh Burkhardt, Rosemary Fraser et al (see, for example, Binns et al 1989 and Fraser 1989).

- Several projects by the Spode Group and the Centre for Innovations in Mathematics Teaching at the University of Exeter, directed by David Burghes (see, for example, Hobbs and Burghes 1989).

- The Mathematics Applicable Project, directed by Christopher Ormell at the University of East Anglia (see Ormell 1982).

(3) From Australia

- The Mathematics in Society Project (MISP), an international project based in Australia, directed by Alan Rogerson (see Rogerson 1986).

(4) From the Netherlands

- The HEWET Project at the OW & OC Institute, University of Utrecht (see De Lange 1987).

(5) From the FR Germany

- The Mathematikunterrichts–Einheiten–Datei (MUED), an association consisting mainly of school teachers (see Böer/Meyer–Lerch/Volk et al 1987/88).

Apart from such larger projects, there are numerous interesting individual contributions of different sorts from all over the world. Again we shall confine ourselves to mentioning some materials published in the eighties, without any claim of giving a complete account.

Firstly, we name two examples of *modelling courses*: Burghes, Huntley and McDonald (1982) and Giordano and Weir (1985). Very helpful are *sourcebooks* and *collections* of applied examples, for example Bushaw et al (1980). Much material can be found in the proceedings of the three International Conferences on the Teaching of Mathematical Modelling and Applications (ICTMA–1,2,3; for references see part II). As particularly regards *mathematical problem solving*, we would like to mention Schoenfield's monography (1985), the extensive conference proceedings edited by Silver (1985), and the proceedings of the problem solving theme group at ICME–5 (Burkhardt et al 1988). Finally, we refer to the two proceedings of the ICMI conference on Mathematics as a Service Subject (Udine 1987), an ICMI Study edited by Howson et al (1988) and an anthology edited by Clements et al (1988).

Many of the materials referred to incorporate the use of *computers* to a substantial degree, in accordance with the apparent trend towards a quantitative and qualitative extension of the use of computers in mathematics instruction, as identified in section 2.4. Therefore it is useful to consider once again the role and impact of computers in mathematics instruction at school and university, with special respect to problem solving, modelling, applications and links to other subjects.

3.3 The role and impact of computers

We shall reflect here only on computers as a *tool* for mathematics instruction, as a *means* for performing numerical or algebraic calculations, or for drawing graphs and pictures, and as an *aid* to creating new teaching methods. We mention only three aspects relevant to our topic.

- The relief from tedious routine makes it possible to *concentrate* better on the *applicational* and *problem–solving processes* (see section 1.1) and thus to advance important process–oriented qualifications with learners.
- Problems can be analysed and understood better by *varying* parameters and *studying* the resulting effects numerically, algebraically or graphically.
- Problems which are inaccessible from a given theoretical basis may be *simulated* numerically or graphically.

As is well known, the existence of powerful tools always has implications not only for methods but for *goals* and *contents* as well. Three apparent aspects in relation to our topic are present.

- Routine computational skills are becoming increasingly devalued, whereas *problem solving abilities* such as building, applying and interpreting models, experimenting or simulating are becoming *revalued*.
- *New types of content* which are particularly *close to applications* can be treated more easily now, eg difference and differential equations or data analysis at the upper secondary level, and statistics, optimisation, dynamical systems or chaos theory at the tertiary level (cf section 3.5).
- It is possible and – because of their increasing relevance in application practice – also necessary to deal with such *applied problems* that *necessitate* the use of *computers* to a considerable extent.

Everything said so far may sound very positive and promising. However, it should be recognised that computers may also entail many kinds of *problems* and *risks* in relation to problem solving, modelling and applications, for instance the following.

- *Arithmetic and geometric skills and abilities* of learners may *atrophy*, though they are still indispensable for problem solving and real world applications.

* The devaluation of routine skills – skills which hitherto have helped students to pass tests and examinations – will make mathematics instruction *more demanding* for all students and too demanding for some of them.

* Paradoxically perhaps, teaching and learning may become even *more remote from real life* than before, because real life may now only enter the classroom through computers, simulations may replace real experiments, computer graphics may serve as substitutes for real objects.

* The use of ready-made software in applied problem solving may put the *emphasis on routine and recipe-like modelling*, thus neglecting essential activities such as reflecting upon the meaning and suitability of concepts and results within a mathematical model.   To put it more succinctly and more generally, *intellectual efforts and activities* of students may be *replaced by mere button pressing*.

* More and more mathematics teachers are becoming interested in *computers instead of* in *problem solving, modelling and applications*, and more and more students are being prevented (or rather – like to be prevented) from reflecting on challenging mathematical problems (pure or applied) by being engaged in technological problems (which would not exist without computers). So, the growing interest in computers and their increasingly easy availability in the classroom may, in some cases, even act to the *detriment* of problem solving, modelling, applications and relations to other subjects in mathematics instruction.

We have no easy patent recipe to offer for solving these problems. Perhaps the most important remedy is a very elementary one: teachers *and* students should become fully *aware* of these problems.   At the same time this would contribute towards one of the vital goals of mathematics instruction (cf section 1.2), namely the acquisition of critical competence in and meta-knowledge of mathematics, its relations to applications and the advantages and risks of its tools.

In the last three sections of this paper, we shall consider *curricular consequences*.   Let us assume, for a given mathematics programme, that it is decided to make problem solving, modelling, applications, or cooperation with other subjects part of the mathematics instruction. What would/should be the consequences of this decision and its implementation for the *organisation and methods* of mathematics instruction, for the *spectrum of topics* in the curriculum, and for *assessment and tests*?

3.4 Consequences for the organisation

In slightly modifying continuation of the categorisation suggested in the Applications and Modelling Theme Group report of ICME-5 (see Lesh et al 1986), the following different *types of basic approaches* to including problem solving, modelling and applications to other areas in mathematics programmes seem to prevail.

(A1)   **The separation approach**.   Problem solving, modelling and applications work are cultivated in separate courses specially devoted to them.   In this way, the pure mathematics courses may remain unaffected.

(A2)   **The two-compartment approach**.   The mathematics programme is divided into two parts.   The first part consists of a usual course in pure mathematics, whereas the second one deals with one or more of the items problem solving, modelling and applications to other areas, utilising mathematics established in the first part or earlier.

(A3)   **The islands approach**.   The mathematics programme is divided into several segments each organised according to the two-compartment approach.

In (A2) and (A3), the closer in time and content the relationship is between pure mathematics sections and subsequent sections concentrating on problem solving, modelling and applications, the more the latter sections tend to assume the character of being traditional *exercises*.

(A4)   **The mixing approach**.   Elements of applications, modelling and problem solving are frequently invoked to assist the introduction of mathematical concepts and so on.   Conversely, new developed mathematical concepts, methods and results are activated towards applicational, modelling or problem situations whenever possible.

(A5)   **The mathematics curriculum integrated approach**.   Here problems, whether mathematical or applicational, come first and the mathematics to deal with them is sought and developed subsequently. In principle the only restriction is that the problems considered lead to mathematics which is relevant to and tractable in the given mathematics curriculum.

(A6)   **The interdisciplinary integrated approach**.   This approach is largely similar to (A5) but differs from it in that this one operates with a full integration between mathematical and extra-mathematical activities within an interdisciplinary framework.

The decision of which (combination of) approach(es) should be adopted for a given mathematics programme depends on a multitude of factors: the arguments for, and the purposes and goals of, problem solving, modelling and applications in mathematics instruction and the characteristics and peculiarities of the educational (sub)system under consideration.

Viewed in an international perspective, the general picture of mathematics programmes that have included problem solving, modelling and applications, seems to be, if painted with a broad brush, the following. In *elementary* mathematics instruction in school the approaches (A3) and (A4) are, for obvious reasons, predominant. This is also the case with secondary school mathematics, but in a few instances approaches (A5) or – more seldom – (A6) can be seen as well. At *tertiary* level the diversity is larger. In mathematics as a service subject, all approaches can be encountered, but probably (A2), (A3) and (A4) are the ones most widely used. Also in courses orientated towards students of mathematical sciences all approaches may occur, but (A1) and (A3) seem to prevail. Approach (A2) is, however, fairly popular too.

3.5 Consequences for the spectrum of topics

The leading questions for this section are the following. Are new mathematical topics becoming relevant for mathematics curricula as a result of the inclusion of problem solving, modelling and applications to other areas? If Yes, which? Are there old topics which could be left out? If Yes, which? Since in this paper we are concerned with mathematics instruction at any educational level, and in all sorts of educational systems exhibiting an abundance of peculiarities and differences, we have to concentrate on rather general matters.

If we begin by looking at *mathematical problem solving* which is focussing on *general* processes rather than being explicitly directed towards extra-mathematical modelling or applications, it seems that the impact on the topics profile of mathematics curricula is not very strong.

If we look at *primary* and *lower secondary* school mathematics, the mathematical topics represented therein are, almost by definition, basic to all mathematical activity, including problem solving, modelling and applications. Therefore, the selection of topics at those levels are *largely unaffected* by the strengthening of problem solving, modelling and applicational components in mathematics instruction. This does not imply, though, that the emphasis put on different elements of the topics remain unaltered.

Next, we turn to *applied* problem solving, modelling and applications at *upper secondary* and *tertiary* levels (see also Pollak 1989). It is often stated, and rightly so, that there is no such thing as applied mathematics because experience has shown us that all mathematics is actually or potentially applicable, if not sooner then later. Yet, in addition to classical calculus, differential equations, numerical analysis, linear algebra, probability and statistics, some *new mathematical topics* have emerged and/or gained momentum in post-elementary mathematics curricula. We need only mention discrete and finite mathematics, difference (and more generally functional) equations, iterations, dynamical systems, chaos, fractals, graphs and networks, optimisation, stochastic processes, stochastic differential equations, control theory. On the other hand, since the total amount of space and time at our disposal for mathematics instruction has not been expanded, the inclusion of problem solving, modelling and applications and related new topics in the mathematics programmes has resulted in a *reduction* of the scope left for *traditional topics.* This is especially true for topics which used to spend much effort on developing cunning formulae to facilitate technical computations, many of which may now be carried out by computers without difficulty. Examples of this are special functions, and special differential equations.

3.6 Consequences for assessment and tests

Before dealing with the generic question of this section, the forms of assessment and tests which are appropriate for a sensible evaluation of activities in problem solving, modelling or applications, to other areas, we should remind ourselves that the issue of assessment and tests has different facets to it.

Firstly, any kind of assessment and testing in relation to mathematics instruction may serve to evaluate either the *students* or the *instruction.* In both cases assessment and tests are *practice*-orientated. However, it may also be that it is not established whether a certain kind of knowledge or skill can be taught and learned *at all.* In this case, issues of assessment and testing are *research–and–development*-orientated.

The second facet is concerned with the *role* of assessment and tests. Is their role to provide *information* to individual students about the quality of their achievements, or is it to provide a basis for *decisions* or *measures* to be taken in relation to the individual, for instance the verdict 'passed' or 'failed', sanctions, awards, privileges and the like? In this facet we also include the question of where to place assessment and tests in the curriculum; should they be continuous, occasional, or only final?

The third and last facet to be mentioned here is the character of assessment and tests. What we mean by this is whether they are to refer to defined *standards*, whether they are to be carried out in *qualitative* or in *quantitative* (scores, grades, marks) terms, and whether they are to be *relative* or *absolute.*

How do the considerations just presented specialise to problem solving, modelling and applications? In *primary* and *lower secondary* mathematics instruction, assessment and testing of pupils' abilities to utilise mathematics in solving applied problems which are neither open nor too complicated have a long tradition in the school systems in most countries. So, in that respect, assessment and tests are mainly practice-orientated. If, however, we consider modelling, open applied problems and more sophisticated mathematical problems, then assessment and tests are rather objects of research and development.

Broadly speaking, the same holds with *upper secondary* and *tertiary* mathematics instruction. There seems to be general agreement amongst those engaged in the field that problem solving, modelling and applications *can* be taught and learned, and that knowledge and skills belonging to them can be reasonably assessed and tested. However, especially if higher order abilities are involved, it is necessary to use forms of assessment and testing which cannot be formalised or standardised very easily.

Altogether, for post–elementary mathematics instruction, assessment and tests in relation to problem solving, modelling and applications are in an experimental stage. Substantial problem solving, modelling and applications abilities are only objects of systematic assessment and testing in a few curricula around the world. There is no doubt that this constitutes a bottleneck to a widespread integration of problem solving, modelling and applications components in mathematics instruction. For, in an examination–based education system, as most educational systems are, instructional components which are not tested on a par with other components tend to occupy marginal positions only.

In conclusion we could say that currently the *role* of assessment and tests is to *inform* students and teachers rather than to provide bases for decisions or measures, and that the *character* of assessment and testing is mostly *qualitative* and *absolute* with no reference to well–defined standards.

REFERENCES

Bell, M. (1983). Materials Available Worldwide for Teaching Applications of Mathematics at the School Level. In *Proceedings of the Fourth International Congress on Mathematical Education* (ed. M. Zweng *et al*). Birkhäuser, Boston, 252–267.

Berry, J. *et al* (eds) (1984). *Teaching and Applying Mathematical Modelling*. Horwood, Chichester.

Berry, J. *et al* (eds) (1986). *Mathematical Modelling Methodology, Models and Micros*. Horwood, Chichester.

Berry, J. *et al* (eds) (1987). *Mathematical Modelling Courses*. Horwood, Chichester.

Binns, B. *et al* (1989). Mathematical Modelling in the School Classroom: Developing Effective Support for Representative Teachers. In *Blum, W. et al* (eds), 136–143.

Blechman, I., Myskis, A. and Panovko, J. (1984). Angewandte Mathematik. *Deutscher Verlag der Wissenschaften,* Berlin–Ost. (Original in Russian: 1976).

Blum, W. (1985). Anwendungsorientierter Mathematikunterricht in der didaktischen Diskussion. In *Mathematische Semesterberichte*, **32**, 195–232.

Blum, W. (1988). Report of Theme Group 6: Mathematics and Other Subjects. In *Proceedings of the Sixth International Congress on Mathematical Education* (eds A. and K. Hirst). János Bolyai Mathematical Society, Budapest, 277–291.

Blum, W. *et al* (eds) (1989). *Applications and Modelling in Learning and Teaching Mathematics*. Horwood, Chichester.

Böer, H., Volk, D. *et al* (1987). Handlungsorientierung. *Mathematik lehren*, 25.

Böer, H., Volk, D. *et al* (1988). Mathematik im Alltag. *Mathematik lehren*, 26.

Burghes, D., Huntley, I., and McDonald, J. (1982). *Applying Mathematics – A Course in Mathematical Modelling*. Horwood, Chichester.

Burkhardt, H. (ed) (1983). *An International Review of Applications in School Mathematics*. ERIC, Ohio.

Burkhardt, H. *et al* (eds) (1988). Problem Solving – A World View. *Proceedings of Problem Solving Theme Group ICME-5*. Shell Centre, Nottingham.

Bushaw, D. *et al* (eds) (1980). *A Sourcebook of Applications of School Mathematics*. NCTM, Reston.

Clements, R. *et al* (eds) (1988). *Selected Papers on the Teaching of Mathematics as a Service Subject*. Springer, Berlin/Heidelberg/New York.

Cross, M., and Moscardini, A. (1985). *Learning the Art of Mathematical Modelling*. Horwood, Chichester.

Fraser, R. (1989). Role of Computers for Application and Modelling: The ITMA Approach. In *Blum, W. et al* (eds), 373–380.

Giordano, F. and Weir, M. (1985). *A First Course in Mathematical Modelling*. Brooks/Cole, Monterey.

Hobbs, D., and Burghes, D. (1989). Enterprising Mathematics: A Cross-Curricular Modular Course for 14–16 year olds. In *Blum, W. et al* (eds), 159–165.

Howson, G., and Wilson, B. (eds) (1986). *School Mathematics in the 1990s*. Cambridge University Press.

Howson, G. *et al* (eds) (1988). *Mathematics as a Service Subject*. Cambridge University Press.

Kaiser, G., Blum, W., and Schober, M. (1982). *Dokumentation ausgewählter Literatur zum anwendungsorientierten Mathematikunterricht*. FIZ, Karlsruhe. (Supplement to appear in 1990).

Kaiser-Messmer, G. (1986). *Anwendungen im Mathematikunterricht*, 1/2. Franzbecker, Bad Salzdetfurth.

De Lange, J. (1987). *Mathematics – Insight and Meaning*. OW & OC, Utrecht.

Lesh, R., Niss, M., and Lee, D. (1986). Report on Theme Group 6: Applications and Modelling. In *Proceedings of the Fifth International Congress on Mathematical Education* (ed Carss, M.). Birkhäuser, Boston, 197–211.

Niss, M. (1987). Applications and Modelling in the Mathematics Curriculum – State and Trends. In *International Journal for Mathematical Education in Science and Technology*, **18**, 487–505.

Niss, M. (1988). Report on Theme Group 3: Problem Solving, Modelling and Applications. In *Proceedings of the Sixth International Congress on Mathematical Education* (eds A. and K. Hirst), Budapest, 237–252.

Niss, M. (1989). Aims and Scope of Applications and Modelling in Mathematics Curricula. In *Blum, W. et al* (eds), 22–31.

Ormell, C. (1982). *The Applicability of Mathematics*. Maths Applicable Group, Norwich.

Pollak, H. (1979). The Interaction between Mathematics and Other School Subjects. In *New Trends in Mathematics Teaching IV* (eds UNESCO). Paris, 232–248.

Pollak, H. (1989). Recent Applications of Mathematics and their Relevance to Teaching. In *Blum, W. et al* (eds), 32–36.

Rogerson, A. (1986). The Mathematics in Society Project: A New Conception of Mathematics. In *International Journal for Mathematical Education in Science and Technology*, 17, 611–616.

Schoenfeld, A. (1985). *Mathematical Problem Solving*. Academic Press, Orlando.

Silver, E. (ed) (1985). *Teaching and Learning Mathematical Problem Solving: Multiple Research Perspectives*. Erlbaumer, Hillsdale.

Sloyer, C. (1989). Contemporary Ideas in Applied Mathematics. In *Blum, W, et al* (eds), 258–261.

Steiner, H. G. (1976). Zur Methodik des mathematisierenden Unterrichts. In *Anwendungsorientierte Mathematik in der Sekundarstufe II* (eds Dörflet, W., Fischer, R.). Heyn, Klagenfurt, 211–245.

Usiskin, Z. (1989). The Sequencing of Applications and Modelling in the University of Chicago School Mathematics Project 7–12 Curriculum. In *Blum, W. et al* (eds), 176–181.

[1] An extended version of this paper is to appear in 1990. This is available as a preprint in *Mathematische Schriften Kassel* 1989 and in *Tekster fra IMFUFA* 1989.

CHAPTER 2

Historical and Epistemological Bases for Modelling and Implications for the Curriculum

U. D'Ambrosio
Campinas University, Brazil

1.   MODELLING AND THE REAL WORLD – EPISTEMOLOGICAL REMARKS

There has been much effort in trying to relate mathematics to other subjects in school. One talks about relationships between mathematics and the real world, mathematics and other school subjects and applications of mathematics. In some circles one brings integrated science into play to show how, through integration, mathematics would play a fundamental role in knitting together various modes of scientific thought (see D'Ambrosio, 1979). In other circles mathematics and science educators face the growing problem of failure in the sciences because students are unable to apply mathematics to elementary situations in the sciences, even if they have shown proficiency in the operative aspects of the necessary mathematical subjects in earlier schooling. When faced with a different situation they are unable to use what they have shown they know (see Hartwig, 1988; the exhaustive bibliography there collected is in itself an important contribution to the theme of this paper). Efforts to close this gap between learning mathematics and using it in other subjects has been a concern of mathematics educators for some time, again trying to knit together mathematics and other school subjects (see D'Ambrosio, 1979). The basic thread in this knitting is techniques of *modelling*. When we mention the important step of applications, we use mathematics as a tool to achieve a certain result in a stage somewhat advanced in coping with the situation we are facing. Identification of the major issues involved has already been done and we have the global picture of the situation. This is even more true when we talk about problem solving. The mere formulation of the problem identifies the main instruments to be put into use for its solution. So the famous remark "A well formulated problem is a half solved problem". Instead, modelling is that stage at which one is faced with a situation in a real context subject to an undefinable number of parameters, some of them even unidentifiable. It provides a focus into the situation and, according to the power of our scientific vision, identifies and selects a number of parameters and builds up a *model* of the real situation. This is the essence of creativity (see D'Ambrosio, 1983 and 1986). As an example, I mention the research, conducted in a slum near Campinas, by Marcelo de Carvalho Borba (1987) as part of his dissertation. He approached a group of young teenagers in the community as they reached the decision to build a soccer field. From the very early steps of identifying a suitable piece of land (done with the model of a plane), through the more difficult stages of clearing the land (making profit from the waste and the woods – commercial arithmetic) and of dimensioning the field (proportions), all was done through the identification of a few necessary parameters and the mathematical manipulation of these parameters.

In talking about real context we are assuming reality as the primary source of our reflections about the situation we are facing and which we try to model, which is nothing but approximating the real situation by what we call *the model*, in which only the parameters which we have identified and selected, according to our capability of dealing with them will delineate the model. Then, with the tools we have and knowing how to use them – this is a major factor in our choice of the parameters – we *work* the model, drawing results, insights, information, indeed learning about the situation, explaining it, coping with the problems which are inbuilt in the situation which usually motivates our approach to it. These are what we have elsewhere called "really real problems" which pose the challenge of solving them, because indeed this means coping with a "situation problem". The diagram in Figure 1 illustrates this behaviour.

This is the very essence of the intelligent inquiry which distinguishes homo sapiens from other species. Hence, modelling is the essential feature of human intellectual behaviour. Although conceding that some people – mathematicians – use certain kinds of tools better, based on manipulation of numbers, measurement, ordering and sorting and inferring, essentially performing what is nowadays called mathematical modelling, we claim that this behaviour is much more general. We refer to D'Ambrosio (1986) for more considerations of an epistemological nature. Let us move on to some historical reflections.

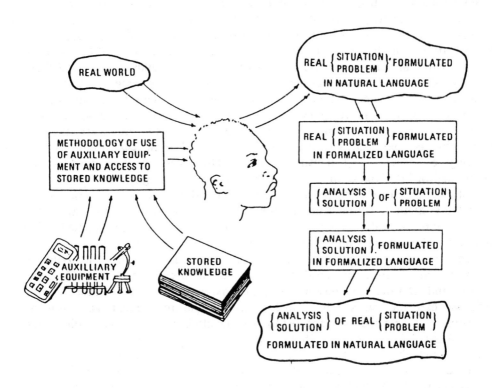

Figure 1

2.   SOME HISTORICAL REFLECTIONS

By mathematics we understand the mode of thought which began to take form in Greece some 2,500 years ago and which was shaped through medieval and renascence Europe into its current forms.   The overall objectives of this mode of thought are, as an etymological analysis would reveal, an art or technique (techne = tics) of understanding, explaining, learning about, coping with, managing, the natural, social and political environment.   Other cultural systems were also looking for their own art or technique of understanding, explaining, learning about, coping with, managing, the natural, social and political environment, and the divinatory nature, hence mysticism, associated with these objectives are again undeniable.   In particular, many of these techniques rely on processes like counting, measuring, sorting, ordering, inferring.   This search, which continued throughout history, has been, and continues to be, the essential motivation of well identified cultural groups for building up corpora of knowledge which came to be called religion, art, philosophy and science. When we say well identified cultural groups we mean a group of people who share common and distinctive civilization characteristics, such as jargon, codes of behaviour, hopes and fears, or summing up, language and culture in their broad sense. We might say "ethnic group" in the broad and modern acceptation of the root *ethos*, which has been in colonial times abusively associated exclusively with race. We call *ethnomathematics* the art or technique of understanding, explaining, learning about, coping with, managing, the natural, social and political environment, relying on processes like counting, measuring, sorting, ordering, inferring which result from well identified cultural groups. A relevant example of cultural bounds on the use of syllogisms, with obvious implications for the conceptualisation of inference, is recorded by Marcia and Robert Ascher (1986), after Scribner. A Kpelle would answer the question "All Kpelle men are rice farmers. Mr Smith is not a rice farmer. Is he a Kpelle man?" with a response which partially says that "If you know a person, if a question comes up about him you are able to answer. But if you do not know the person, if a question comes about him, it's hard for you to answer." Similarly, the question "If A or B then C, given not A and B, is C true?" would be answered "If A then Q, if B then not Q, if not Q then not C, given not A and B, the conclusion is not C". As the authors observe, the logic is fine but the specific world view of the respondent got into the reasoning. Consequently, the −*tics* he develops for his *mathema* would be distinct.

3. INTERDISCIPLINARITY

In this theoretical framework, the disciplinary distinction based on specific codes, paradigms and even symbols come in a much later stage, both from the point of view of individual cognitive behaviour and of the historic evolution of ideas. In the schools, new subjects, more appealing to modern concerns, such as statistics, systems analysis, chaos, catastrophes, programming, decision theory and different applications to other fields of knowledge have to find space throughout the curriculum. Some of these, for example statistics, although fully established as a branch of mathematics and even a discipline in itself for close to a century, have not, as yet, broke through the conservatism of the mathematics curriculum, not to mention computer science. Some others are not yet fully established mathematical fields, they are more like mathematics in the making, which may be an additional reason to bring them to the schools, since they are in need of fresh new directions and new ideas. Why not bring to the students the feeling that we are building−up knowledge, just like a research mathematician is a builder of knowledge, not a mere utilizer of accumulated knowledge. Without any doubt this would enhance creativity, which is the least cared for among the reasons for keeping mathematics in schools.

Let us again address the specific theme of the relation between mathematics and other school subjects. In a growing pattern, knowledge becomes the subject of quantification. Sociology, political sciences, psychology and even literary fields are joining economics and other subjects traditionally placed among the humanities in claiming credibility through their use of mathematical models. In some cases, it is as yet no more than the use of mathematical jargon. But this is the first step towards adopting a mathematical way of thought in dealing with their subjects. This is probably one of the characteristics of our era. Credibility is associated with precision, demonstrability – in a certain mathematical style – and rigour, and the mere use of mathematical jargon brings scientific status to some subjects. Clearly, this is an indicator of how influential mathematics is in modern society. Mathematical thinking has acquired unprecedented prestige. Maybe this is the main reason why mathematics is kept with such intensity as the major school subject. This will prolong well into the 21st century, and calls for interdisciplinary activities and even popularization of mathematics. Of course, this can be achieved only if mathematicians adopt an attitude of tolerance with respect to background and even style of thought. Popular mathematics will be needed and this will ask for going against mystifying ("involving in mystery or obscurity" in The Random House Dictionary 1966) the main areas of mathematics. For this a different, less authoritarian, classroom posture is needed. Clearly evaluation must change from the pass/fail style and all its variants (grades, exams, tests) to a creative enhancing exercise.

Interdisciplinarity demands more flexibility in dealing with different areas and different styles of mathematical thought. This will require a broader and highly diversified basic knowledge, acquirable only through modern, advanced encyclopaedic techniques, something in the line of the stored knowledge given in the figure above. Although reluctance towards this form of knowledge acquisition prevails among mathematicians, its prevalent use among competent scholars, mathematics–users from other fields, it will be imposed to mathematicians themselves. We are already witnessing indications of this. The ability of non–mathematicians to use in an efficient and competent way, rather sophisticated advanced mathematics, is a fact. We now see an undergraduate student in physics dealing comfortably with fiber bundles, and biologists going deep into nonlinear differential systems. Both move into these fields, cope with rather complex mathematical techniques and obtain new results with a background which is considered, from the traditional standards of mathematicians, weak and unacceptable. The handling of mathematical results and techniques, acquired in what I call an encyclopaedic way by non–mathematicians, will probably force the revision of mathematics contents in schools, bringing rather advanced subjects to the lower levels of schooling. It is already a fact that even advanced PhD students are not prepared to follow advances outside their narrow area of dissertation. It is clear that the prevailing linear, cumulative style of designing syllabi for mathematics courses does not satisfy the requirement of more and more advanced knowledge, which must be imparted earlier and faster. In other words, we have to bring what is today considered advanced mathematics to earlier and broader stages of education. Scientists and especially mathematicians have to be better informed of recent advances in mathematics. In other words, they have to have more mathematical culture.

REFERENCES

Ascher, M., and Ascher, R. (1986).   Ethnomathematics.   *History of Science*, v.24, pp.125–144.

Borba, Marcelo de Carvalho (1987).   *Um estudo de Ethnomatematics: sua incorporacao na elaboracao de uma proposta pedagogica para o Nucleo–Escola da Favela de Vila Nogueria e Sao Quirino.*   Dissertation, Universidade Estadual Paulista (UNESP), Rio Caro.

D'Ambrosio, U. (1983).   Non–formal Educational Modules and the Development of Creativity.   In L. D. Gomez, P. (ed.), *Creativity and Teaching of Science*, San Jose, Costa Rica: CONICIT/Asociacion INTERCIENCIA, pp.79–90.

D'Ambrosio, U. (1986).   Reflexions sur le mode de pensée occidental et sur la science et l'education.   In Rapport final sur le Colloque de Venise: *La Science face aux confins de la connaissance: le prologue de notre passe culturel*.   Paris: UNESCO, pp.141–152.

D'Ambrosio, U. (1979).   Strategies for a closer relationship of mathematics with the other sciences in education.   In UNESCO/CTS/ICMI/IDM Conference on *Co–operation Between Science Teachers and Mathematics Teachers*.   Bielefeld: IDM, pp.27–34.

Hartwig, Dacio R. (1988).   *Uma Estrutura para as Operacoes Fatoriais e a Tendencia na Utilizacao de Formulas Matematicas: Um Estudo Exploratori*.   Doctoral dissertation, Faculdade de Educacao, Universidade de Sao Paulo.

CHAPTER 3

Statistics and the Problem of Empiricism

J. Evans
Middlesex Polytechnic, UK

ABSTRACT
Statistics can be made comprehensible within the teaching situation by viewing it as a set of methods concerned with the assessment and development of theories and models. This raises the question of the relationship between *theories* and *data*, which is seen in a one-sided way by empiricist views of science. An alternative two-sided view of the theory/data relationship is discussed, which presents several opportunities. First, it can illuminate the teaching of statistics. Second, it provides a convenient practical checklist for students to use, when they are critically evaluating a research report in their other subject, or when they are designing their own research projects.

1. INTRODUCTION

This chapter aims to respond to a challenge set down by Ubi d'Ambrosio at ICME 6: "... we are therefore also facing an historico-epistemological problem when we discuss the relations of mathematics and other subjects, and interdisciplinarity" (Hirst and Hirst, 1988, p 282). The chapter focusses especially on the epistemological aspects of the problem, with particular reference to my experiences teaching mathematics, statistics and research methods, to undergraduates and post-graduates in the social sciences in several British institutions, including Middlesex Polytechnic and the Open University; discussions with a more historical and social focus may be found elsewhere (eg Irvine, Miles and Evans, 1979; Radical Statistics Education Group, 1982).

2.   THE PROBLEM OF KNOWLEDGE IN EMPIRICAL DISCIPLINES

This chapter starts from a somewhat different perspective from that used in, for example, the model conception quoted by Blum in Hirst and Hirst, 1988, p 278.   There, the focus is on the relationship between, on the one hand, *mathematics*, including mathematical models, and on the other, the *real world*, including real models based on other disciplines.

Here, I focus on the relationship between *models* (the theoretical) and *data* (the empirical) based on several principles.   First, the real world cannot be apprehended directly, but only in a mediated way, through our ideas, models, theories.   I understand these to be basically theories from various empirical disciplines, in various forms, sometimes in mathematised form.   Examples of these other disciplines would be economics, psychology, sociology, biology, physics.   Mathematicians see these other disciplines as a source of applications; these other disciplines see us as providing help in assessing and developing their theories and models.

Second, to understand the real world, we undertake theorising and modelling in various stages:

1.   *problems* are *formulated*, through seeking to understand the world on the basis of our theories and models;
2.   using these theories and models, we are able to *act* in the world; a by-product of this is that our theories are constantly assessed and developed.

In certain situations, some called "research" or "science", however, the *main* aim of the activity is to assess our theories and models; in this case, step (2) becomes a cycle in which mathematical and statistical ideas are especially useful, as follows:

2A.   a *research* study is designed, and
2B.   *data* is produced, through work in the real world;
3.   the data is *analysed*;
4.   the analysis is *interpreted* so as to develop our theories and models.

Steps 1, 2A, 2B, 3 and 4 can be seen as stages in the process of empirical research, or, put another way, in the modelling process (eg Open University, 1983, pp 4-6).

3.   THE PITFALL OF EMPIRICISM: A ONE-SIDED VIEW

An ever-present danger in the teaching of mathematics and statistics in the context of other subjects is that of *empiricism*.

Empiricism is that view of epistemology which insists (or more correctly, attempts to insist) that:

- all knowledge comes from data, or observation;
- theory in these other disciplines develops through the collection of data, and induction, or the making of generalisations, from data;
- progress in science will come from collecting ever-bigger samples and developing more sophisticated techniques for generalising from them.   (See, for example, Doyal and Harris, 1986, Chapter 1 for

further discussion of empiricism).

Thus it can be seen that, in philosophical terms, empiricism recommends a *one-sided*, excessively data-based approach to science and theory development, in contrast to the *two-sided* approach illustrated in Figure 1 below, which balances attention between theory/models and data. This means that fully-developed scientific work needs to emphasise theoretical development, acknowledging the origin and evolving meanings of the concepts current in the discipline, as well as emphasising data production and analysis.

A pedagogic problem related to empiricism is that of an excessive emphasis on the mathematical/statistical issues, at the expense of the other subject's concerns. Thus in terms of *purpose* and *organisational framework* (cf Hirst and Hirst, 1988, pp 279-80), this discussion is focussed on purpose (a) to provide learners with knowledge and abilities concerning other subjects, with mathematics/statistics in a service role, and envisages attempts to work within an organisational framework (b) where mathematics/statistics may be taught as part of, and integrated with, other subjects. Further, the discussion here should be valid, whether the overall aims of the students' education are vocational, or liberal (to do with general education) (Note 1), and whether the students are in tertiary (University, Polytechnic, etc) or secondary education (Note 2).

4.   AN ILLUSTRATION

As an example of this two-sided approach involving both theory and data, let us take the following theoretical statement, from the disciplines of mathematics education, and psychology.

"Women express maths anxiety more than men, at tertiary level".

Consider this first as a theoretical statement. Those working within the other subjects might want to consider a number of issues concerning theoretical development; for example, why, amongst a whole range of possible research questions, this one has been considered interesting, and has engaged the time and resources of a considerable number of researchers, over the last 20 years or so, especially in the US. And so on.

If we wish to assess this statement, or to model the relationship, or simply to describe some aspects of it, we might consider

(a)   what *concepts* it focusses on: here, clearly, gender (women/men) on the one hand, and "maths anxiety" on the other;

(b)   what type of *relationship* is suggested – causal or otherwise: here it may be intended as causal, but most theorists would wish to consider alternative explanations to innate ones (see eg Evans, 1987); and

(c)   what is the *scope* of the statement, or the *population* of institutions and/or individuals referred to: here the statement appears to refer to all students (or all students studying mathematics?) at tertiary level in all (or all "advanced industrial"?) countries.

Though some clarifications are necessary, we can see from this illustration that *theoretical statements* are made up of CONCEPTS, related (often) in CAUSAL terms, for a POPULATION of individuals (and/or institutions, countries) etc.

If we now consider producing *data* that would be useful in assessing this theoretical statement, we find that:

(a)   we would need to specify measures or *indicators* for the concepts mentioned above: it would appear to be easy for gender in principle (Note 3), whereas there is much debate on how to measure "maths anxiety" in a valid way (Evans 1987);

(b)   even if we observed a *correlation* between gender and expressed maths anxiety (or, put another way, gender differences in the latter), we cannot be sure that the relationship is causal (rather than due to an alternative explanation – based on, say, educational or social factors) – without further controls;

(c)   we usually can study only a *sample* (or subset) of the population in which we are interested.

Thus we can say that *data* are based on INDICATORS, related via CORRELATIONS, for a SAMPLE.

5.   AVOIDING EMPIRICISM: A TWO–SIDED APPROACH

We can put together these three of the *theoretical* and the *empirical*, as in Figure 1(a) below

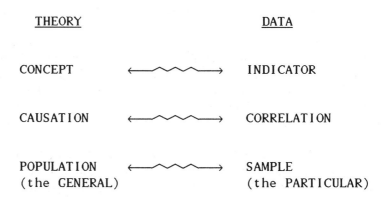

Figure 1(a)   Three Facets of the THEORY–DATA Relationship

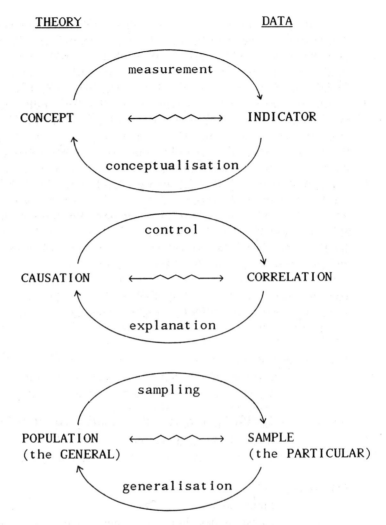

Figure 1(b)   Three Facets of the THEORY–DATA Relationship
                     and Consequent Problems for Research Design and
                     Analysis

A number of points may be made about this scheme.

1.   The basic form of the scheme can be easily grasped by first–year
     undergraduates, or by interested secondary students, and may even
     be familiar since the three facets may appear in common–sense
     argument.   Only basic numeracy is needed to elaborate the
     discussion.

2.   The scheme in Figure 1(a) portrays an essentially two–sided
     relationship: the two sides interpenetrate yet it is useful to be able
     to think about them as (analytically) distinct.   (Note 4).

3.   We can give names to the processes involved in moving from the theoretical to the empirical, and back again; see Figure 1(b). For example, we could say roughly that *measurement* (producing (a) valid indicator(s) for a given concept), *control* (specifying a correlation (or set of correlations) that will effectively test a certain causal relation), and *sampling* (choosing an appropriate sample for the population of interest) are problems for the design stage of research (Note 5). On the other hand, the three labels attached to the leftward-pointing arrows come up at the problem – formulation stage, and are revisited at the analysis and interpretation stages; viz *conceptualisation* (what concepts should be used to organise the data), *explanation* (accounts of why what was observed was observed), and *generalisation* (attempts to apply our theories in other contexts). This separation, however, would be too neat: many decisions at the design stage depend on how we plan to analyse and interpret our data, and of course vice-versa.

4.   The six problems shown in Figure 1(b) relate roughly to familiar topics in most statistics (and research methods) courses in the social sciences as follows:

i(a)   measurement
-    representing variation numerically/coding
-    types of measurement scales (categories, rankings, interval scale, ratio scale)
-    validity and reliability of indicators
-    methods of data collection
-    descriptive statistics (for summarising the whole sample)

i(b) conceptualisation
-    factor analysis
(most issues here are theoretical/conceptual, rather than technical)

ii(a) controls
-    experimental or survey design
-    use of probability-based methods (eg   random allocation to treatment groups)
-    comparative method (in ethnography and history)

ii(b) explanation
-    analysis of cross-tabulations
-    coefficients for correlation and partial correlation
-    multiple regression
-    analysis of variance

iii(a) sampling
-         random and non-probability sampling methods

iii(b) generalisation
-         statistical   inference   ie   confidence   intervals   and
significance tests.

5.   There   is   clearly   some   overlap   between   the   three   facets;   for
example, probability methods are used both for random allocation to
groups   (ie   control)   and   for   random   sampling.   However,   a
separation of the facets is useful for clarifying certain confusions;
see   for   example   Kish's   comments   (1959,   p   330   ff)   on   certain
criticisms of tests of significance in sociology.

NOTES

1.   For a discussion of the differing objectives of vocational, liberal and
academic education, see Robinson 1968, Chapter 4.

2.   For a description of a set of materials for teaching statistics in
various   niches   in   the   secondary   curriculum,   see   Schools   Council
1980.

3.   However, it can often be difficult to assign a gender to every
student, for example, in a national testing programme where only
names have been collected.

4.   In empiricist approaches to the social sciences, the right side of
these facets would dominate the left hand side or the facets would
be collapsed altogether.   Thus, we have operationalism, which urges
that a CONCEPT should be defined by the measurement operations
necessary to produce scores on the INDICATOR; for example, some
psychologists want to *define* intelligence as a person's score on a
Stanford-Binet   IQ   test   administered   under   certain   conditions.
Similarly, many present-day social scientists present causal analysis
(as derived from J. S. Mills' method of agreement and differences)
for *inductively* deriving CAUSAL relations from CORRELATIONS
only; see eg Cole (1972).   Finally, many have hoped to find rules
of *induction* for generalising straight from particular samples to
general theories.   For discussion of fallacies in the hopes of
operationalism and inductivism, see Doyal and Harris (1986).

5.   The labels attached to the rightward-pointing arrows are those used
by Kish (1959).

REFERENCES

Cole, S. (1972). *The Sociological Method.* Chicago: Markham.

Doyal, L. and Harris, R. (1986). *Empiricism Explanation and Rationality: an introduction to the Philosophy of the Social Sciences.* London: Routledge and Keegan Paul.

Evans, J. (1979). Causation and Control, Part 1 of *Block 3A: Research Design. DE304: Research Methods in Education and the Social Sciences.* Milton Keynes: Open University.

Evans, J. (1987). Anxiety and Performance in Practical Maths at Tertiary Level: a report of research in progress. pp 92–98 in Bergeron, J. C, Herscovics, N and Kieran, C. (eds) *Psychology of Mathematics Education (PME XI).* Proceedings of the Eleventh International Conference, Montreal, July.

Hirst, A. and K. (eds) (1988). *Proceedings of the Sixth International Congress on Mathematical Education.* Budapest: János Bolyai Math Society.

Irvine, J., Miles, I. and Evans, J. (eds) (1979). *Demystifying Social Statistics.* London: Pluto Press.

Kish, L. (1959). Some Statistical Problems in Research Design. *Am Social Rev.* **24**, June, 328–338; reprinted in Tufte, E. R. (ed) (1970) *The Quantitative Analysis of Social Problems*, 391–406. Addision Wesley.

Robinson, E. (1968). *The New Polytechnics: The Peoples' Universities.* Hammondsworth, UK: Penguin.

MDST242 Course Team (1983). Unit AO: Introduction. *MDST242: Statistics in Society.* Milton Keynes: Open University.

Radical Statistics Education Group (1982). *Reading between the Numbers: a critical guide to educational research.* London: BSSRS (25 Horsell Road, London, N7).

Schools Council (1980). *Teaching Statistics 11–16: statistics in your world.* Slough, UK: W Foulsham.

CHAPTER 4

The Identification and Teaching of Mathematical Modelling Skills

A. O. Moscardini
Sunderland Polytechnic, UK

1. INTRODUCTION

An increasing number of companies and governments are beginning to use modelling and simulation as cheap and reliable ways of assessing, optimising and controlling processes and systems and this is producing a rising demand for modellers (Andrews 1983). In response to this demand, courses including the term Mathematical Modelling are proliferating in Europe and America. One must question what and how such courses teach the student (Burghes 1986). The term Mathematical Modelling has many interpretations but will be used to describe the more restrictive process by which one arrives at a mathematical formulation of an initial problem as opposed to describing the total modelling process.

2. THE MATHEMATICIAN'S VIEW OF MATHEMATICAL MODELLING

The view of mathematics that is taught is often different from the practice of mathematics.

This statement is one possible reason for the unpopularity of mathematics in the classroom. The teaching of modelling attempts to correct this by teaching what is practised. There are two principal schools of thought, the platonist and formalist, on the nature of mathematics and most mathematicians can be placed in one of them (Davis 1983). The platonists believe that mathematics consists of timeless, eternal laws that are waiting to be discovered. These laws exist independently of mankind. The formalist's creed is a self contained, consistent body of knowledge invented by mankind and used by mankind to describe the world. Most mathematicians would say that they are platonists but actually are practising formalists. It is the formalist viewpoint that lends itself more readily to a modelling process.

A good modeller will possess certain skills. It is the intention of this paper to try and identify what precisely these skills are and examine different ways of teaching them. Before this can be done, the stages in the modelling process must be identified and explained.

3.   THE MODELLING PROCESS

3.1 Problem analysis
In this stage, the modeller investigates the background to the problem and how the results are going to be used. Every model is built for a purpose and it is essential to be clear about this purpose. One must research and discover all the information that is available and especially the form and accuracy of any data or parameters. Planning a model accurate to eight decimal places is only meaningful if the data can be estimated to the same accuracy. Knowledge elicitation techniques as described here, require general skills such as researching papers and books, questioning, understanding and selective listening. These are not regarded as mathematical techniques. The result of this stage is normally a concise set of aims and objectives for the problem.

3.2 Problem modelling
For want of a better term, this stage requires mathematical modelling skills. It is at this stage that the mathematical description of the problem is obtained. The identification and teaching of these skills form the main thrust of this paper.

3.3 Analysis of model
In true industrial and commerical modelling, the equations are usually extremely complicated and require much solution time. It is therefore important to be reasonably sure that the equations are correct before one begins to solve them. Of course, one can never be certain that this is the case but there are certain tests that can be done, such as looking at steady states, phase-planes analysis, looking at special cases, extreme values and parameter estimation. The skills involved are quite specialised and would be termed mathematical.

3.4 Solution of model
The skills at this stage are definitely mathematical. A suitable technique must be chosen and applied correctly to obtain a solution to the problem. The skills required are mathematical knowledge, judgement, manipulative and/or programming ability.

3.5 Validation

A true validation is often difficult and sometimes even impossible. Often there are no facts to compare results with, just one's own preconceived notions. This stage involves experimental work and will usually involve sophisticated statistical techniques. This can be the most highly mathematical stage of them all.

4. DEVELOPMENT OF MODELLING SKILLS

The process of doing mathematics and the process of problem solving are different.

It is becoming apparent that two different types of skill are necessary to complete the modelling process which require two different modes of thinking. De Bono (1980) makes the distinction between vertical thinking and lateral thinking. Vertical thinking is thinking in a linear fashion where every step is a logical progression from the previous one. It is intimately connected with the YES/NO system of logic where one must be correct at each step and if an idea is not right then it is wrong. Lateral thinking is best illustrated by the analogy of traffic moving along a major highway. It is sometimes better or quicker to reach a destination by taking a sideroad and bypassing the highway, ie it is sometimes easier to reach a solution by addressing the problem from a non-standard approach. This type of thinking does not need the strict logic of the YES/NO system and De Bono has postulated a new system as an alternative. Stages 3.4 and 3.5 which rest heavily on strict mathematical correctness may be said to need vertical thinking but the other stages certainly require the more lateral approach. Thinking occurs in two stages:- Perception and processing. For the more traditional mathematical problems, little perception is needed as the direction of the solution is obvious. In a more realistic situation, however, a wrong perception at the initial stages can result in a tremendous amount of extra work. In fact if the initial perception is totally wrong then no amount of correct mathematics will rescue the situation. Lateral thinking is connected with the perception stage while vertical thinking is connected to the processing stage. These two types of thinking seem to correspond to the two types of skills required in a good modeller. There is no attempt to promote one type above another. Both skills are needed and both types of thinking are used.

4.1 Mathematical modelling skills

It is presumed that the first stage has been completed, ie the problem has been analysed, aims and objectives have been produced. The skill is now to translate these aims and objectives into mathematical form. The practising mathematician when asked what he would do in this situation would probably write down the general equation that suited the problem and then decide which of the terms he should keep and which he could safely discard. What qualities enable him to do this? Some possibilities are:- experience of similar situations (models) over many years, the fact that he is an expert in that particular area or an intuitive feel for the correct mathematics.

An interesting experiment would be to take a mathematician out of his specialised area and observe how he derives the equations. This would illustrate the difference between the applied mathematician and the modeller. The teacher's problem is that normally one is dealing with students who are not experts in any particular area, do not have experience to fall back on and do not yet have fully developed mathematical intuition. The problem is to devise teaching schemes where all these qualities can be fostered. Two general skills involved are the identification and representation of the relevant variables and the establishing of the relationships between them.

4.1.1 *Variable identification*
The identification of variables corresponds to the expert's judgement of what terms are important in his equation. A common method employed in teaching is the feature list. Here all factors that could remotely affect the problem are written down in the form of a list. Then each factor is examined in turn and an assumption is made about it. It could be that the assumption is to ignore this factor or that it has some special feature such as it is constant etc. The obvious drawback to this method is how does one decide which features to ignore and which to include. At the early teaching stages, the advice is to keep the model as simple as possible so as to only include the terms thought to be absolutely essential. This is in fact good advice to novice modellers: keep the initial model as simple as possible in the knowledge that, later, terms can be introduced gradually, each time making the model more realistic. It can be compared to a sculptor whose initial effort is to get the rough shape of his creation right before worrying about the details. The advantage of the feature list is a teaching one. It forces the student to think carefully about his assumptions and provides a definite stage in the methodology for this to happen.

The problem of deciding which variables are important is not a trivial one but the obvious gaps in the feature list approach led a colleague, Don Prior (1986), to try an alternative approach using systemic methods and ending with the model in the form:

Outputs are a function of inputs.

This is a very useful form of a model as it can now be developed in many ways, ie if a mathematical model is wanted then the problem now is to find the form of the functional dependence. There are still many weaknesses to this approach but one advantage is that it generates a tremendous amout of discussion. Our experience is that students really get involved and are learning the essential skills of refinement, discernment, honing, and judgement.

4.1.2 *Formulation of equations*

Students find this stage most difficult. There are no established methods of formulation techniques so two methods used at Sunderland Polytechnic are discussed. The first can be loosely described as using the INPUT–OUTPUT principle. For each variable, the following word equation is postulated:

```
Amount of variable   =  Amount of variable at previous time
at present time
                     +  Amount created in interval

                     -  Amount destroyed in interval

                     +  Net inflow into system
```

This leads through a difference equation to the general form:

$$\frac{dV}{dt} = f(V) \qquad \text{where } V(t) \text{ is the variable}$$

This progression of word equation – difference equation – differential equation is a useful one but has obvious limitations as not all models are of this type. But the philosophy behind the approach is to demonstrate that the mathematical equation is just a mathematician's way of desribing the problem. Thus the student is coaxed into writing down the relationship in words first, then translating the words into the language of mathematics. Another method in use at Sunderland Polytechnic is causal loop modelling (Goodman 1980) where the controlling mechanisms are identified. This is the equivalent of the stage described in section 4.1.1. Causal loop diagrams can easily be translated into a system dynamics diagram which can be solved by a software package such as DYNAMO or for translation into difference equations. Again one will argue that there are many models that do not fit into this structure. The point to remember is that one is trying to give the student some sort of methodology to enable him to practice the skill of formulating equations. Practice can be obtained from problems that obviously lead to differential equations (which form the majority) and then other formulations can be studied by examining models that use different mathematics such as Leslie matrices.

Formulation of equations is very difficult and there exists no recognised methodology for teaching it.

5.   IMPLICATIONS FOR THE TEACHING OF MODELLING

It is a moot point whether skills can be taught but good teaching can certainly foster or nurture skills.   The dictionary definition of a skill is a practiced ability so the principal feature of a modelling class is to provide the most suitable environment for the student, not to practise mathematics (this is done in mathematics classes), but to practise the process of describing a problem by using the language of mathematics.

As this is difficult and is hardly ever correct first time round, it is most important to create an atmosphere where the student is not afraid of being wrong or of making a mistake.   To do this the lecturer must abandon his expert status and become more of a guide or consultant. In the lateral thinking tradition, mistakes are regarded as useful stepping stones to better solutions.   Splitting the students into small groups helps to create this atmosphere.   In a group everyone has a say and the student learns to listen and accept criticism from his peers.   The groups then report back to the class at frequent intervals which gives practice in presentational skills.   The students are encouraged to think freely with no restrictions except the knowledge that anything they say will be discussed.   After a group has worked on a problem it is quite in order for the lecturer to discuss the group's work in front of the class and if possible to follow this with perhaps one of the methods described above. However, even then it must be stressed that the lecturer's answer, whilst it is a model of the problem, it is not the only one (Moscardini 1986).

There is no one correct solution to a modelling problem.

It is still instructive, however, to study published or working models. One problem here is that the modelling stages are very rarely stated and most modelling papers are more concerned with the process after the equations have been formulated.   The only published accounts of modelling that I have found are contained in the biographies of famous scientists such as Copernicus, Newton, Einstein and Bohr.

It is clear that different skills are being practised here to mathematical skills and in the same way different assessment procedures are needed.   In Britain the formal examination of three hours is accepted as THE true way to test anything.   Modellers, possibly more than academics, would agree that, as in industry, deadlines have to be met and therefore one must learn to think under pressure, but nowhere in real life is a scientist, modeller or engineer required to think in such a small interval of time as three hours.   What needs to be tested in a modelling course is the student's ability to produce a model, criticise it and react to the criticisms by improving the model.   This must be done within a deadline but this deadline could be a number of weeks not hours.   The student may have consulted his peers and may have used some of their ideas (correctly attributed).

There are various ways of examining this: a series of assignments, open book all day examination or oral examination on the students work.

It is essential that in the early stages of planning a modelling course, the assessment procedures are fully discussed (Moscardini 1984). The three hour exam has one very distinct advantage: it is a very efficient way of examining in the sense that it uses the least resources. This argument must be resisted because when dealing with modelling one could be efficiently examining totally the wrong aspects.

Modelling needs special assessment procedures.

Sunderland Polytechnic is active in research in these areas and the author would be interested in further communication on any points.

REFERENCES

Andrews, J. G. (1983). Academic Applied Mathematics: A view from Industry. *IMA Bulletin*, **Vol 19**, No 3.

Burghes, D. (1986). Mathematical Modelling – Are We Heading in the Right Direction. *Mathematical Modelling Methodology, Models and Micros* (eds) Berry, J. S. et al. Ellis Horwood.

Davis, P. J. and Hersh, R. (1983). *The Mathematical Experience.* Penguin Books.

De Bono, E. (1980). *The Use of Lateral Thinking.* Pelican Books.

Prior, D. E. (1986). A New Approach to Model Formulation. *Mathematical Modelling Methodology, Models and Micros.* (eds) Berry, J. S. et al. Ellis Horwood.

Goodman, M. R. (1980). Study Notes in System Dynamics. MIT Press.

Moscardini, A. O. (1984). The use of Small Projects in a Part-time Course. *Mathematical Modelling Courses.* Ellis Horwood.

Moscardini, A. O. (1984). Issues Involved in the Design of a Mathematical Modelling Course. *Mathematical Modelling in Science and Technology.* Pergamon Press.

CHAPTER 5

Some Theoretical Aspects of Descriptive Mathematical Models in Economic and Management Sciences

Günther Ossimitz
Institut für Mathematik, Universität für Bildungswissenschaften Klagenfurt, Austria

ABSTRACT

This paper contains seven theses about the role of mathematics in contemporary economics and management science. Although mathematics is a widespread tool for describing economic and business structures, its usefulness is restricted by the arbitrariness of the basic economic quantification act, which really is an act of measuring. Thus the application of mathematics works well only in rather small well–structured parts of our empirical and economic world. The didactic consequences of the theses are discussed briefly at the end of the corresponding sections.

1. MATHEMATICS IN ECONOMIC AND MANAGEMENT SCIENCES IS MOSTLY DESCRIPTIVE MATHEMATICS OR ELEMENTARY ARITHMETIC

Mathematical terms and concepts like functions, graphs, equations are frequently used to *describe* economic models (see for instance Lipsey et al 1987). For example let us consider a production process. In economics it is often described as a *function* with some (independent) input *variables* and usually one (dependent) output variable, which represents the value of the production–function for the given input. Mathematical concepts (variables, functions, slope of functions etc) are used to describe economic notions. In this context usually a strange duality between quantity and quality arises.   Although the mathematical concepts are in general quantitative, they are often used in a qualitative manner.   A production function for example is usually just sketched graphically without any scales on the axes, or described in a form like

Output = f(Labour,Capital)

On the other hand mathematical calculations in economics and business are used frequently only at the level of *elementary arithmetic*.   Invoices, book-keeping and many other common business calculations consist mostly of elementary arithmetical operations.   Most calculations concerning interest or amortisation funds can be regarded as repetitive elementary arithmetic.   Complex mathematical algorithms or computations (eg for optimising some commercial strategies) are relatively rare both in economic education and practical business.

As another example let me consider the payment for human labour. The mathematical operations involved in calculating the exact wages are pure arithmetic.   On the other hand the single figure of personal income has also a descriptive dimension: it can be used for determining welfare, social status and so on.   It can also be aggregated over many people in order to describe the distribution of income in a society or to compare the income of different groups or nations.

In the context of (business-orientated) mathematical education I would like to stress two aspects:
- business mathematics is the most natural and most important application of elementary arithmetic and thus the *primary education of elementary mathematics* should be taught with a strong *impact on business applications*;
- on higher levels a fair grasp of the *descriptive power* of mathematical concepts (like functions, graphs, tables, etc; see for example Fischer 1984) is desirable.

2.   THE   FUNDAMENTAL   ACT   OF   MATHEMATISATION   IN   ECONOMICS AND BUSINESS IS THE ACT OF MEASUREMENT

What is meant here by measurement?   Let me define *measurement* as a *systematic quantitative description* of *qualitative structures*.   In this definition the term systematic implies that
- certain measurement rules are observed:
- qualitative relations among the objects being measured are still valid for the measured numbers.   If Object A is heavier than Object B we expect that after weighing both A and B the weight of A should be a bigger number than the weight of B (no matter how the measurement is accomplished in detail).

The most common example of economic measurement is the determination of the value of goods or services, measured in monetary units.   In the case of payment for human labour, any wages actually include a kind of quantitative evaluation (measurement) of the work being done. Sometimes the value of some work (or goods) depends upon two prior measures: the *quantity* of the work (good) – measured in some appropriate unit – is multiplied by the *value per unit* (eg the wages per hour). In other cases the monetary value (or price) is determined directly, eg if employer and employee agree in a free contract about the salary to be paid. This latter form of measurement is an example of *fundamental measurement*, whereas the former case (quantity multiplied by value per unit) is an example of *derived measurement*, since the measured value is derived from other values. Since any empirical figure relies upon quantification, it is evident that economic mathematization always involves, at least conceptually, some form of measurement.

Mathematical measurement theory (eg Roberts 1979) regards fundamental measurement as a kind of mapping which preserves certain relations like order, additivity and suchlike. This view of measurement is a *mathematical idealisation* in itself. It does not take into account that measurement is more than just a special kind of structure–preserving mapping. In my opinion one of the central features of any measurement is the following.

3. FUNDAMENTAL MEASUREMENT TRANSFORMS QUALITATIVE MATTER INTO QUANTITATIVE STRUCTURES

This thesis implies that measurement is also something that links two essentially different worlds: the world of qualitative, complex and mostly diffuse structures is mapped via measurement into the world of well structured, straightforward mathematics. It is just the difference between the work you do and the money you get for it. Even the difference between the work being done and the working time gives us a glimpse of the radical transformation taking place when something is being measured. On the one hand we have working conditions, the degree of satisfaction with the work, the special skills required for that particular work etc. All these qualitative aspects may differ considerably from case to case. On the other hand we have the quantitative figures of wages or working time, which can easily be compared among different persons. They can also be averaged over different groups. They can even be changed without changing the quality of work at all. It makes a great difference whether you compare two jobs or just the payment for these jobs. Another aspect is that the earnings from regular work can be added together with other forms of income, such as from investments. Thus we can conclude that measurement does not simply *reduce complexity*, it also introduces a new, *quantitative dimension*, which allows new operations.

I also want to stress that this *transformation from quality to quantity* implies that any act of measurement is not as trivial as it might seem at first sight. It requires a thorough didactic reflection to let pupils and students grasp that fact and its consequences for using and interpreting empirical figures or data.

4. ECONOMIC MEASUREMENT IS TO A GREAT EXTENT ARBITRARY REGARDING BOTH WHAT IS MEASURED AND HOW IT IS MEASURED

The arbitrariness of *what is being measured* becomes clear when we consider some examples. When we go shopping it is rather arbitrary that we have to pay just for the goods themselves, whereas the entrance into the shop, the service and, in many cases, the wrapping of the articles are free. At a trade-fair usually an entrance fee has to be paid. Another example is the field of taxation. It is rather arbitrary that human labour (at least in Austria and many other countries) is fined with taxes whereas the use of natural resources (like air, water, energy etc) and the production of litter are free from extra taxes. Many forms of human work are paid for, some (like the work of housewives or the task of educating children) are not.

The arbitrariness of *how something is being measured* can be shown easily. Just let us consider how prices or wages can be fixed:
- they might be fixed by an official authority (government etc);
- they might be the result of some commercial cost calculation;
- they might be the result of an individual agreement between purchaser and buyer or employer and employee;
- a price might be the result of competition in a market.

Of course the above thesis does not imply that everybody is free to choose how to measure something in economic or business life; actually considerable organisational, legislative and scientific effort is undertaken to establish and maintain general standards of measurement. An income tax law for instance is a set of rules on how to measure income. The exact formulations of the law, how income is determined (and which expenses are tax-deductible and which are not) are rather arbitrary. Nevertheless all citizens are obliged to declare their income according to the national income tax law.

The arbitrariness of measurement is closely related to the abstractness of mathematical terms and concepts. Mathematical concepts are, to a very high degree, free of empirical contexts and this gives us much freedom in measurement and quantification. The same set of natural numbers, the same concept of a real function etc fits to an uncountable number of different applications. This aspect of freedom and arbitrariness when applying mathematics contrasts sharply with the strictness of mathematical rules and might seem unfamiliar to our common view of mathematics.

In mathematical education we usually deal with mathematical laws and rules, we hardly reflect upon the freedom and arbitrariness of measurements and quantifications we need for applying mathematics in business and every day life. This issue is a major challenge for teaching applied mathematics, where considerable didactic research and reflection has still to be undertaken.

5. QUANTITATIVE DESCRIPTIONS OF ECONOMIC SITUATIONS ARE RATHER ARBITRARY, AS FAR AS THEY ARE BASED UPON ANY ACT OF MEASUREMENT

This thesis is a simple consequence of the former thesis and gives an idea of why economic models are often rather formal without many empirical connections. It also gives a clue why in practical business only crude arithmetic is common.

Since the statement of the last thesis sounds very negative, we should ask, of what use is mathematics in economics and business?

6. ECONOMIC QUANTIFICATION AND MATHEMATISATION WORKS WELL IN A LIMITED, WELL-STRUCTURED CONTEXT, WHERE A CERTAIN AMOUNT OF QUANTIFICATION IS ALREADY GIVEN. IT USUALLY FAILS TO BUILD UP A PROPER COMPREHENSIVE REPRESENTATION OF THE COMPLEXITIES OF OUR QUALITATIVE WORLD, ESPECIALLY WHEN NON-QUANTIFIABLE ASPECTS PLAY AN IMPORTANT ROLE

When buying several articles in a shop a simple addition of the prices works perfectly for determining how much to pay. But if we want to determine the utility of these goods for our personal demand and would like to *measure* the utility, things usually become complicated. Let us suppose that somebody has bought some bread and butter and wishes to spread the butter on the bread. In this case the utility of the bread might depend upon the availability of butter and probably also on the availability of some substitute for the butter. On the other hand the utility of the butter might depend upon whether the refrigerator works or not. Thus the simple evaluation of utility might become complicated and – if actually carried out – rather arbitrary.

The payment of wages works well in practical life because usually fixed salary schedules, general wage agreements and other prescriptions are already given. It would be far more complicated and in many cases practically impossible to take into account the whole complexity of the work or to attempt to define precisely the value of the work done.

I think that it is also an important task of application-oriented mathematical education to deal with the pragmatic setting of practical measurement and mathematization. The main problem of applying mathematics is not only to find the appropriate mathematical tools and rules but also the freedom and arbitrariness of measurement and quantification and its consequences for handling data. Such investigations have to be undertaken in the context of the actual application and I think that the insights from reasoning about the quantification process might differ considerably from case to case.

7. THE ROLE OF MATHEMATICS IN ECONOMICS AND BUSINESS
    IS RATHER SIMILAR TO THE ROLE OF CHEMISTRY IN
    MEDICINE. BOTH ARE SUCCESSFUL ON A SPECIALISED
    AND WELL STRUCTURED LEVEL BUT THEY SHOULD NOT
    BE REGARDED AS THE ONLY AND UNIQUE TOOL FOR
    ATTAINING GLOBAL AND UNIVERSAL TRUTH IN THE
    FIELDS TO WHICH THEY ARE APPLIED

Many people believe that a mathematical description of something
increases the validity and precision of reasoning. We tend to believe
that something that is determined mathematically must be exact just
because mathematics does not allow any imprecision. This view does not
take into account that there are virtually no mathematical restrictions on
how to use mathematics in our empirical world and that mathematical
modelling implies the formulation of measurement rules and other
arbitrary decisions. A questionable model and questionable quantifications
will lead to questionable results – despite the fact that all mathematical
calculations within the model are mathematically precise.

This leads me to the final moral: any result of a modelling process
depends explicitly or implicitly upon arbitrary (free) decisions or
agreements. Whenever we deal with figures measuring some aspects of
our empirical world, we should be conscious about the freedom and
arbitrariness of the quantification process. This will help to prevent
much of the abuse of mathematics in application contexts. And this is a
task for application–oriented mathematical education on all levels.

REFERENCES

Davis, P. J. and Hersh, R. (1986). Descartes' Dream: The World
According to Mathematics. Brighton. The Harvester Press.

Fischer, R. (1984). Offene Mathematik und Visualisierung. *mathematica
didactica* 7, 139–160.

Lipsey, R. G., Steiner, P. O. and Purvis, D. D. (1987).   Economics.
New York. Harper and Row.

Roberts, F. S. (1979).   Measurement Theory with Applications to
Decision–making, Utility, and the Social Sciences.   Reading, Mass.
Addison–Wesley.

CHAPTER 6

The Human and Social Context for Problem Solving, Modelling and Applications

A. Rogerson
Hawthorn, Australia

ABSTRACT
Problem solving and modelling has already provoked much debate along the lines of what is a problem and what do we mean by modelling? The general philosophical answer to such questions, I believe, has already been clarified in Wittgenstein (1967, 1968) but the debate is still useful if it provides, as Polya's classic work did, new examples of problem solving, and hence of problem solving methodology (Polya 1957 and Duncan 1979). The overall message of this paper is that for us to fully understand both the theory and practice of problem solving, modelling and applications, we should look *outwards*, to the historical development of applied mathematics, to the rich fields of modelling in the sciences, to the many problem solving casestudies in the history of mathematics, and (most of all) to the complex socio-cultural contexts from which all real world applications are derived (Rogerson 1982).

1.   THE HERITAGE OF APPLIED MATHEMATICS
How far does the present interest in problem solving, modelling and applications represent a radical shift in emphasis, and how far is it merely a continuation of previous trends in what used to be called applied mathematics?   One way of answering this question is to look in some detail at the historical development of mathematics curricula and texts with particular reference to the prevailing social and cultural conditions.   For example, one of the major characteristics of industrial society in nineteenth century Britain was its insistence on utilitarian

arithmetic in schools to meet the need for trained clerks in commerce and industry. A standard work of that time, Nesbit's Practical Arithmetic, for example, would seem to modern eyes to contain an incongruous mix of abstract theory taught in a traditional manner, alongside hundreds of problems, some of the most practical or utilitarian sort. On page 133, for example, we find this typical question: "A gentleman has ordered a rectangular court–yard to be paved which measures 45 feet 9 inches in front and 32 feet 6 inches broad. A foot path 5 feet 6 inches in breadth, leading to the front door of his house, is to be laid with Portland–stone, at 3s 4d per square foot, and the rest with Purbeck–stone at 2s 3d per square foot; what will be the expense of the whole?"

What is interesting about many such examples in Nesbit's book is the insertion of so much background and incidental detail on matters which nowadays would only be found in technical or practical studies: information on trees and timber (including two poems lyricising the oak tree!), stone cutting, much commercial accounting, elementary science and so on. From this we may argue that mathematics education at that time *was* highly applicable and certainly more closely related to the socio–cultural context than nowadays. Early in the twentieth century, however, it is clear that Victorian practicality had largely given way to books such as Nunn (1913) which was almost entirely made up of artifical problems – journeys of two snails, baths being filled with water from two taps with different velocities (later the baths would have three holes in them for the water to flow out!), and so on.

Later still in the 1930s, there began a strong reaction both to this artificiality and also to the earlier Victorian utilitarianism, summed up in Ballard (1931) who sarcastically criticised not so much the genuine practicality of the earlier arithmetic, but precisely the kind of pseudo–problems and armchair artificiality that this had degenerated into: "It was believed that children hated multiplying 5247 by 78 but if they were allowed to write the word 'apples' or 'elephants' or 'bales of cotton' after the 5247 they would regard the multiplying as a privilege and a joy. It was weariness of the flesh to reduce five hundredweights to ounces, but to reduce five hundredweights of coal to ounces was as cheery a business as reducing them to ashes". It would be historically misleading, however, to assume that the viewpoint of Ballard was the prevalent opinion of mathematics educators of that time. On the contrary (then as now) authors went on writing, and publishers went on publishing, texts full of artificial and unmotivating pseudo–real problems.

2.   CONTEMPORARY APPLIED MATHEMATICS

If we now turn to the more recent history and development of school curricula, at least in the UK, it is clear that, for pupils aged 16–18, traditional applied mathematics has predominantly meant *mechanics*, a long established and elaborated model of forces, velocities, bodies etc including statics, dynamics, kinematics and (less frequently) hydrostatics, hydrodynamics and electricity.   The first observation about applied mathematics at this level is that it has lost touch with *university* courses in mechanics which use vectors, matrices and partial differentiation in the development of the general theory and Hamilton's Equations.   This would be bad enough if traditional applied mathematics at school was nevertheless motivating for pupils in applying mathematics to the real world, but is it?

A close examination of many traditional mechanics texts of the past twenty years reveals a plethora of questions about ships steaming north against currents flowing east, men walking at three mph and five mph towards a crossroads, batsmen striking cricket balls at angles of 35° to the vertical, frictionless pulleys, men in lifts holding spring balances, and so on.   None of these examples are *real* (or even realistic), they are artificial exercises created to practise a deductive model which is itself a collection of loosely connected and sometimes contradictory principles: Newton's laws of motion, the observation of energy and momentum, the so called laws of friction...   Even though traditional applied syllabi, and some texts, have now updated their content, the examples and exercises are still largely stereotyped and artificial.

So much for traditional applied mathematics, but have the reforms of modern mathematics, particularly in Britain, improved this situation as far as applied mathematics in the upper secondary school is concerned? There have been significant improvements in content – the introduction of vector and matrix methods, statistical and probabilistic models and new areas of application including electricity.   These have often been introduced in an open, relevant and apparently applicable manner, but all too soon it is confessed that reality is too complicated and we must fall back on simplified (ie unreal) examples and exemplars.   As far as the SMP (School Mathematics Project) A-level course in the UK is concerned, there is little genuinely open investigation, nor any extended discussion of real life, only routinised and artificial problems elaborating a deductive and closed model.   A typical example occurs in SMP (1968) where a chapter begins by considering how we might work out the volume of a cooling tower, wine glass, bath or rugby ball.   While it is stimulating, for example, to discuss the cooling tower, we soon discover that we are not able to find its volume because reality is too complicated and the chapter continues with Further Calculus Techniques and Applications – its title!

While attempts to apply mathematics to electricity at school level seem to be unreal and symbol-juggling, there is an opportunity for real life problems to be covered in probability and statistics.   The extent to which this is done varies enormously, from treatments that are almost entirely pure mathematics, through to modelling casestudies in which statistical and probability ideas can actually be discovered to help solve real life problems.   We may note in passing that even though mechanics, and more recently electricity and statistics/probability are formally considered as applied areas of mathematics, virtually every branch of mathematics is applicable, including calculus, trigonometry, geometry and arithmetic, as is clear from a detailed examination of their use in the sciences. This has already been fully discussed in Rogerson (1979, 1981 and 1989) so will not be considered here, except to say that it represents a vitally important aspect of modelling and problem solving.

3.   THE RELEVANCE OF THE HISTORY OF MATHEMATICS

The seminal work of Kuhn (1962) and Lakatos (1976) has already radically changed the conception that scientists and mathematicians have of their respective disciplines. Kuhn's training as a historian of science made him aware that history is not the passive accumulation of past knowledge or facts, but is rather an open-ended and dynamic attempt to recreate the past (Collingwood 1961). In as much as history (and hence the history of mathematics) is a *human* activity involving the subjective interpretation of documents and other primary sources, it must therefore always be culturally relative (Rogerson 1982). Although there are still many texts that present the history of mathematics as a lifeless authoritarian body of knowledge that has somehow accreted in a simplistic, evolutionary fashion, there is also a growing number of books and projects that recognise the history of mathematics as a dynamic, interpretative and essentially *human* activity, for example Davis (1981). This humanistic view has proved to be *didactically* useful since it provides us with a series of detailed casestudies of modelling and problem solving, using imaginative recreations or rational reconstructions of historical events (Carr 1961).

For example, the traditional deductive/abstract method of introducing complex numbers is to start from a definition and deduce the properties, using examples and exercises to reinforce the manipulative skills required. The alternative is first to pose the following problem: why is it that polynomial equations of degree n sometimes have n roots and sometimes fewer than n roots? By discussion and guided discovery we can lead students to imaginatively reinvent the complex numbers. It is vital that we did not give the students the solution to the initial problem too soon. This approach was found to be successful because it explained why complex numbers were invented and what purposes they serve.

The success of this and similar examples is reminiscent of the fruitful ideas of problem posing and generative themes expounded in Freire (1971). This experience has been repeated in a variety of other contexts: an introduction to mechanics (Lewis 1971), electrical circuits (Merlane 1971), the vector product (Rogerson 1980) and to selected themes in the SMP 7-13 and MISP projects. It is clear from this experience that we can use the history of mathematics for a host of relevant and motivating modelling and problem-solving situations involving Babylonian and Greek astronomy, ancient Egyptian agriculture and architecture, Islamic art and geometry, and so on (Rogerson 1983). To do this effectively, however, we must regard the history of mathematics as a human, creative, and therefore fallible development, and, as far as possible, consult and use only the primary sources of more enlightened rational reconstructions (Elton 1961).

4. THE SOCIO-CULTURAL CONTEXT

Why in the past, despite its obvious relevance, has the social and cultural context been so often neglected in mathematics education, for example in the work of Piaget and many of his followers?   Piaget's training as a biologist almost certainly led him astray, as his earlier work indicates, in the simplistic assumption that techniques of analysis of plant growth were applicable to the mental development of children!   Indeed, the whole application of experimental/control models on long term human development, using the vast apparatus, of parametric statistics with its sampling, estimation and hypothesis testing, is misguided if it does not, or cannot, take into account the effect of relevant social and cultural contexts either between individual children or in longitudinal studies with the same children.

While the holistic nature of the social and cultural environment makes it virtually impossible to carry out research in the reductionist mode of most scientific methodologies, there is a viable alternative offered by the research methods practised in the *arts* subjects, rather than the sciences.   These techniques include participant/observer research, piecemeal analysis involving the subjective balancing and weighing of evidence and (most of all) diagnosis and improvement through the introduction of new, creative and therefore human solutions.   During the past 20 years a number of national and international initiatives have attempted to do this, using integrated themes from the socio-cultural context for modelling and problem solving: for example, the intensive work undertaken in the UK from 1961 onwards by the School Mathematics Project (SMP) and, in particular, by the SMP 7-13 Project (1972-1979).   The final two years published materials of SMP 7-13 becomes a kind of proving ground for these new ideas, and for innovative experiments in a problem-solving child-centred thematic approach.   This involved presenting mathematics not as a skeleton list of skills and concepts but rather as a collection of real themes which implicitly contain the mathematics (Rogerson 1986).

At the same time, other writers in Canada and the USA, Italy, Spain, Poland and the UK, often stimulated by the pattern of primary school work, were producing secondary school themes based on the real life of pupils and/or their socio-cultural context.   For example, from 1976 onwards at the University of Genova in Italy, a group of writers had written and tested a three year mathematics/science course based on local historical and economics themes (Rogerson 1983).   Much of this work is being translated into English for use in the Mathematics in Society Project (MISP).   The strong emphasis on modelling and problem-solving in projects such as MISP mirrors work in other subjects in the school curriculum, notably the New Geography, the New Biology (Rogerson 1979 and 1981) and innovative projects such as the Science in Society Project and Spode Group materials in the UK, and the Mathematics at Work and CAM projects in Australia.

Since 1980, the impact of this new perspective on mathematics educators throughout the world has been considerable. Many of them are now placing increasing emphasis on the role of society in the teaching of mathematics and the use of the cultural context as a potent source of examples for real-world modelling (Harris 1980). Those who attended the 1984 ICME-5 congress in Adelaide, Australia will recall the emphasis that Ubi D'Ambrosio placed on ethno-mathematics (ie the socio-cultural context) in his opening plenary address. Four years later Mathematics and Society had even become the theme for a fifth day special at the 1988 ICME-6 congress in Budapest, Hungary. While this attention is welcome (if not overdue) it is hoped that it will not be merely another catchprase or bandwagon for the ambitious, but rather a real and enduring initiative which will eventually restore the socio-cultural context to the central position it deserves in modelling, problem solving and applications of mathematics.

REFERENCES

Ballard, P. (1931). Teaching the Essentials of Arithmetic.

Carr, E. H. (1961). What is History? Penguin.

Collingwood, R. G. (1961). The Idea of History. OUP.

Davis, P. J and Hersh, R. (1981). The Mathematical Experience. Penguin.

Duncan, B. (1979). Some Thoughts on Teaching to Develop Problem Solving Skills. ACT Maths Centre.

Elton, G. R. (1961). The Practice of History. Fontana.

Freire, P. (1974). Pedagogy of the Oppressed.

Harris, P. (1980). Measurement in Tribal Aboriginal Communities. NT Dept of Education.

Kuhn, T. S. (1972). The Structure of Scientific Revolutions. Phoenix.

Lakatos, I. (1976). Proofs and Refutations. CUP.

Lewis, P. G. T. (ed) (1971). Vectors and Mechanics. SMP Further Mathematics Series, CUP.

Merlane, G. (ed) (1971). Differential Equations and Circuits. SMP Further Mathematics Series, CUP.

Nunn, P. (1913). Exercises in Algebra. Longmans.

Polya, G. (1957). How to Solve it. Princeton University Press.

Rogerson, A. (1979). Mathematics and Statistics in Relation to Science. One chapter in *Teaching and Learning Strategies in University Science*. Cardiff University Press.

Rogerson, A. (1980). Teaching the Vector Product Backwards. *The Mathematical Journal of Meerut University*. India.

Rogerson, A. (ed) (1981). Cooperation between Mathematics and Science Teachers (6 booklets). CTS/ICSU/UNESCO.

Rogerson, A. (1982). La Matematica, La Scienza e La Realta. Monograph No 2. Cagliari University, Italy.

Rogerson, A. (1983). Mathematics in Society, The Real Way to Apply Mathematics. MISP Report No 1.

Rogerson, A. (1986). The Mathematics in Society Project: A new conception of Mathematics. *International Journal for Mathematical Education in Science and Technology*, **Vol 17, No 5.**

Rogerson, A. (1989). Mathematical Modelling in the Sciences. In Blum, W. et al (eds) *Applications and Modelling in Learning and Teaching Mathematics*. Horwood, Chichester (UK).

SMP (1978). Revised Advanced Mathematics Book 3. CUP.

Wittgenstein, L. (1967). Remarks on the Foundations of Mathematics. Blackwell.

Wittgenstein, L. (1968). Philosophical Investigations. Blackwell.

CHAPTER 7

Children's City

Designed and built by children in an Architectural Mathematical Laboratory for Cubic City Planning (Video Series for Children Grades 1–5)

M. Anker
United Nations International School and Teachers College, Columbia University, USA

1. ABSTRACT

Architectural Math Drama is a series of plays about social studies issues described through mathematics. The weekly Integrated Day starts with a play inviting the audience to solve a problem through evaluating and redesigning a modular scale model. The plays are about a family moving from one inadequate apartment building to another until they finally move to Children's City. Here the open plan with moveable walls and terraces allow the flexibility necessary to satisfy their changing needs.

Mathematics is used to describe, measure, and calculate the various sets of furniture and building materials and to compare their dimensions to those of the corresponding full sizes surrounding the students in the classrooms and at home. Mathematical operations are used to find the quantities needed to assemble models from room interiors to residential highrise buildings.

The toy people on stage in the modular dioramas formulate the issues, propose solutions and apply their math skills throughout the open-ended dialogues, leading the audience towards the answers without giving them away. Colour cubes on top of blueprints and worksheets add the third dimension, map out concepts graphically and eventually let us assemble the mathematical landscape necessary for overview and holistic problem solving.

2.   PROGRAMME

Since 1983 mathematics has been taught with scale models of modular homes and housing projects to grades one to five at The Centre for the Education of the Gifted, Teachers College, Columbia University and at the United Nations International School in regular, after school and summer programmes.

The popular block corner in kindergarden classrooms has long provided hands–on experiences and a concrete foundation for class–discussions leading into various subject areas. This cross disciplinary approach has been extended to elementary grades by expanding the block corner into an Architectural Mathematics Laboratory. The task here is to assemble metric scale models of classrooms and homes for miniature plastic people illustrating traditional social studies themes starting with the basic privacy and social needs of individuals and families. The open models resemble dolls houses and modular dioramas and the furniture inside represents the various needs and activities of the family. The models are assembled with dm plastic or cardboard building units, grouped in different sets depending on the stage of the construction process.

The students roleplay characters such as factory owner, truck driver, construction worker as well as the family members who move into the home in question. They use mathematics to calculate the quantities of units at these different stages and to measure how realistic the dimensions of the models are. Two cm interlocking colour cubes and cardboard modules are used as furniture units.

At the second grade level the social studies theme is the neighbourhood or local community. Mathematics skills are applied along with map skills within the social studies context. Simplified blueprints and maps of the school and its surroundings are used. The third dimension is again added, with the same colour cubes now representing room units and colours showing main room functions.

Small rooms are built with sets of 1, 2 or 3 cubes of the same colour. Large rooms are assembled with 4, 5 or 6 cubes. Together the room sets, each of a different colour, make up a whole apartment. Rectangular or square apartments are also made of styrofoam flats. Since families increasingly live in city apartments today, students need an overview of common patterns in the organisation of larger buildings and projects. Apartments are grouped and stacked to demonstrate these patterns.

In the third grade the focus is on the town, with its municipal buildings and public services. In order to map out this larger area the scale is reduced further and the colour cube now represents a stack of apartments or highrise units. In the higher elementary grades the emphasis shifts to cardboard models of more complex geometrical shapes.

3.   IMPLEMENTATION

The program is based on a 800 page manuscript of illustrated worksheets and gameboards, plus eight sets of blueprints for each grade level to introduce framework and techniques.   However, reference must continually be made to the actual physical environment of the particular school.   It is therefore necessary that a team of teachers collect and draw up simplified local blueprints.

The same team should also make videotapes of successful classes as examples of how to use the open-ended modular units.   The initial blueprints are only springboards and examples of patterns and combinations, while the final goal is for students to come up with their own unique solutions to the issues of housing.

The more ambitious goal for the team is the production of video programmes described in the sequence mentioned below.   One day each week is designated as an Integrated Day and a double-programme introduces the two sets of activities to be carried out and offered as optional weekend activities for the students.   Scripts of 11 such videos have been tested out in classrooms.

Each video starts with the characters discussing and demonstrating the practical task of arranging interiors or assembling whole buildings. Methods of planning and testing solutions are shown, including common mistakes, and viewers are invited to apply the demonstrated techniques on their own after the video to search for their own personal solutions. There are two ways of doing this – in the lab with plastic equipment and at home with folded cardboard.

The mathematical content centers on spatial thinking and the four basic operations in connection with the metric system drawn up and manipulated on 3D-coordinate grids.   The investigations of dimensions and calculations of statistics are related to the children's own home and school environment.   The social studies content stresses the basic privacy and social needs mapped out with furniture and community facilities and outdoor spaces.

American building traditions are extended to include European, Latin-American and unorthodox ideas, to inspire students to combine existing patterns in new ways.

4.   OBJECTIVES

The major aims of the program are to:
1.   describe mathematically the immediate home and school environment of the student;
2.   provide for concrete and colourful experiences through toy like mathematical equipment and building modules ideal for elementary grades;
3.   use attractive, self-explanatory worksheets with illustrations of cubic and modular housing, helping teachers as well as students to focus as the pictures clarify and simplify concepts and sequences;

4. give researchers an opportunity to observe directly the steps and obstacles in the child's thought–process as s/he builds or rearranges manipulatives;
5. bridge the gap between the abstract paper and pencil world of mathematics and children's natural interest in building blocks and hands–on explorations;
6. integrate mathematics with social studies, drama and language arts bringing mathematics out of its traditional isolation;
7. offer a stage for dramatic scenes from familiar home problems;
8. include built–in evaluation, as completion of manual tasks proves mastery at the same time as it motivates and meets creative needs;
9. make maximum use of minimum manipulatives, reducing the expense and problems of organising and setting up a lab, and providing storage;
10. prove the need for mathematical thinking and tools for the purpose of understanding the environment through the reconstruction of it;
11. provide a modular and cubic mathematical landscape as a concrete foundation and reference point to which to attach abstract concepts, assisting communication and retention;
12. allow numbers and specific details to be seen in a total but simplified context, helping holistic thinkers usually overwhelmed and alienated by details and large numbers.

5. SEQUENCE OF SCENES AND RELATED TOPICS

The main characters are – a boy aged eight, his sister aged ten, their parents, a construction worker Big Hurry, his boss Slow Joe, Building Inspector Sharp, Uncle Grouch. Other characters are – Storekeeper, Mrs Smith a neighour – Jim her son.

5.1 The furniture store

Scene: The storefront of Do–It–Yourself Furniture, Unlimited.
The family is assembling furniture cubes and modules to make their own combinations for their new apartment. They calculate the number of pieces they need, the space the pieces take up in the van and the prices compared to standard type of furniture.

Topics: Sets, length, width, height, prices, tax and discount.

5.2 The apartment

Scene: A small open–plan apartment with bathroom and staircase doors in the background. The kitchen fixtures are also seen in the back of the open space. Against one wall stands a stack of movable walls to be put in place later.
The family is gathered around a small model of the apartment

representing Dad's idea of how to place the walls and furniture.   We hear arguments for and against this solution.   All family members have their own plan and we see how they gradually combine all these ideas into one new solution satisfying most needs.

Topics: Sets of wall and floor units, area, coordinate grids, metric scales.

5.3   The classroom

Scene: An open-plan school where a stack of moveable walls are to be placed along with rolling cabinets, bookcases and other storage units.
    Parents and teachers discuss placement of walls and the furniture. They make projections of the number of students in future classes, and calculate the prices of new sets of furniture.

Topics: Sets of furniture and students, match between height of students and furniture, volume and bills for heating and air-conditioning, area and carpeting prices.

5.4   The after-school centre and sports ground

Scene: A large open-plan space in the Community Centre for the neighbourhood children.
    The discussion among the youngsters is about the dimensions of the space needed for exercise equipment and indoor games such as table tennis, pool, squash and basketball.   Similarly charts are drawn up of the outdoor sports grounds for baseball, football and soccer.   The size of a swimming pool is also discussed.

Topics: Dimensions of rectangular and circular shapes outlining sports fields, diameter and perimeter, volume of pool, game scoring, weights on exercise machines.

5.5   The storefront

Scene: Inside the shop: Kid's Book and Music Exchange.
    The children are starting their own store in the community centre and discuss how to set up and run the business, including taking and paying back bank loans.

Topics: Book-keeping, expenses and profit sharing, interests and percentages, instalments and bank loans.

5.6   The housing cluster

Scene: The office of Joe's Speedy Construction Company.
    The architectural model under discussion is a four-storey apartment

building for a dozen families and a dozen couples or singles. The small cooperative has shared indoor and outdoor spaces, especially for children and the elderly. The protected outside play areas are to be walled in and supervised. Ways of dividing the costs for the shared arrangements are being discussed. Maximum sunlight on the terraces is also considered.

Topics: Sets of building units, quantities per truckload, time for transporation, and construction, rents and building maintenance, family sizes and life cycle, angles of sunrays during the different seasons.

5.7 The housing project

Scene: The office of Joe's Speedy Construction Company.
The full cast is gathered around the styrofoam model of a housing project named The City for Children. The quantities needed for each type of apartment, for each floor, and each cluster of buildings are found with a calculator. Dimensions of streets, driveways and parking spaces are tested with miniature cars. Sunlight on terraces at different times of the day and year is demonstrated with a moveable architects lamp. The needs of shops and public services are discussed.

Topics: Population density and statistics, sets of people, cars and buildings, dimensions of cars, streets and pedestrian areas and paths, sets of building materials, ratio between shops and public facilities and their uses, shadows cast by tall buildings.

6. EXAMPLE OF A TOPIC

Joe's Speedy Construction Company is about to build a block of Railroad apartments. Big Hurry and Slow Joe discuss the number of truckloads needed of walls, windows, doors and floor squares. A model apartment is built and the number of units in each set found. The quantities are multiplied by the number of apartments in the block, illustrating continuous addition of same number. The number of room and floor units is also found by placing the assembled apartment building model inside the upfolded 3D grid and multiplying length, width and height. The units are stacked on a toy truck to establish the number of units per truckload. The next question is how many truck loads are needed from the prefab factory to build the block, showing division through continuous subtraction of the same number.

7. CONTEXT : ENVIRONMENTAL MAPPING AND SOCIAL DESIGN

Architectural Math Drama is a way of proving that abstract mathematics can be used as a tool to describe our physical urban environment and help us reconstruct it. It goes beyond accepting status quo and provides the tools to improve upon existing housing patterns through a nine step creative process named Environmental Mapping and Social Design. However, only two out of nine steps in the skill sequence involve maths. They are Metric Investigations and Reality Reference. Architectural Math is consequently only a small part of an elementary technique for evaluating Urban Design, but it enables us to bridge the gap between the abstract and isolated world of mathematics, so hard to grasp for young children, and social studies and its related disciplines, through the concrete manipulations of mathematical landscapes of Environmental Mapping and Social Design (cf Fig 1).

REFERENCES

Alexander, C. and associates (1977). *A Pattern Language: Towns, Buildings, Construction.* Centre for Environmental Structure, Berkeley, University of California. Oxford University Press. Main framework for programme.

Appleyard, D. (1981). *Livable Streets.* University of California Press.

Hatch, R. (ed) (1984). *The Scope of Social Architecture.* New Jersey Institute of Technology. Van Nostrand Reinhold Company.

Newman, O. (1972). *Defensible Space: Crime Prevention Through Urban Design.* New York. Collier Books.

Sherwood, R. (1978). *Modern Housing Prototypes.* Harvard University Press.

Sommer, R. (1983). *Social Design: Creating Buildings with People in Mind.* Englewood Cliffs, New Jersey. Prentice-Hall.

Untermann, R. and Small, R. (1977). *Site Planning for Cluster Housing.* University of Washington, Seattle. Van Nostrand Reinhold Company.

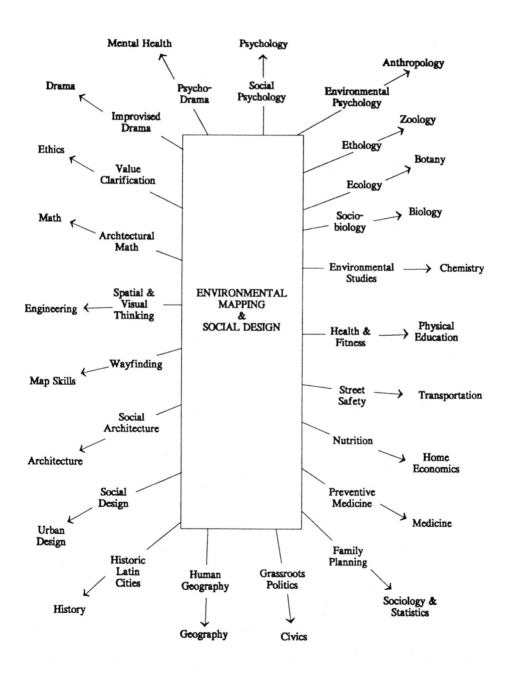

Figure 1

CHAPTER 8

Mathematics, Maori Language and Culture

A. J. C. Begg
Department of Education, New Zealand

ABSTRACT
The Curriculum Development Division is investigating ways to help improve the mathematics learning of Maori students. These methods are based on language, culture, and self-esteem.

The teachers in these classes are wrestling with the following questions:
- should they teach traditional mathematics and merely translate the language into Maori?
- should the teaching start with the mathematics from Maoridom?
- should the teaching reflect the more holistic approach to education that the Maori community traditionally had?

This paper briefly considers these and related questions and looks at the thematic approach which links some of the mathematics with the Maori language themes and which is at present being trialled.

1. BACKGROUND

The *tangata whenua* (indigenous people) of New Zealand are Maori but most people in the country are of British origin and since the colonisation of New Zealand the language of education and commerce has been English. The Maori people are doing what they can to ensure that their language and culture are preserved and within education their is pressure to give parents a choice regarding whether their children are taught in English or in Maori.

The majority of New Zealand schools contain a significant number of Maori students. Europeans make up 79% of the secondary school population and Maori 17%, while Pacific Islanders constitute only 3%

and Asian 1%.   When a bilingual class for Maori students is not practical or desired, then the mathematics (content, applications, and teaching methods) should at least reflect the bi-cultural nature of New Zealand society.

Some people claim that, with the large numbers of immigrants from other countries (in particular the South Pacific and Asia), New Zealand should be multicultural.   However it is the Maori culture that is in danger of being lost in New Zealand and which is not preserved in any other part of the world, while Pacific Island cultures will survive in the Pacific.   Thus the need is seen to stress the bi-cultural nature of the country rather than multiculturalism.

2.   A CLASSIFICATION OF TEACHING

In classifying the range of teaching methods currently being used in high school mathematics classes with Maori students, four types of teaching can be identified, namely:

Traditional (English-style) Teaching
Bicultural Teaching
Bilingual Teaching
Holistic Teaching.

Each of these can be considered more fully.

2.1 Traditional Teaching

The traditional mathematics teaching that nearly all New Zealanders, Maori and *Pakehas* (non-Maori), experience uses teaching methods that were common to English-type classrooms some ten years ago. Mathematics is treated as culture-free and responses are based on the experiences and needs of non-Maori students.   Traditional teaching is concerned with the subject rather than building the self-esteem of each student.

2.2 Bicultural Teaching

In the bicultural classroom greater sensitivity is given to Maori customs. One example of this is accepting that looking down rather than making eye contact is showing respect.

Some applications of mathematics would be set in a Maori context. For example, reading a scale drawing might use the plan of a *marae* (an enclosed ground) with a *wharenui* (meeting house) and a *whare kai* (a building for cooking and dining) rather than a more traditional plan of a piece of land with a house and garage.

Some examples might be chosen taking the mathematics from aspects of Maori language, mathematics or culture, for example:

(a) When introducing measurement, we might use a Maori unit of 1 *maro* rather than informal units for an arm span (or fathom).
(b) When analysing transformations we could look at *kowhaiwhai* designs (rafter patterns).

Figure 1

2.3 Bilingual Teaching

In bilingual teaching situations it is assumed that bicultural features will be incorporated and that the Maori language will be used whenever possible. Before mathematics teaching can proceed using the Maori language, a vocabulary list must be compiled for the mathematical terms that are needed. Such a list is at present being compiled.

The important question is whether to teach European mathematics through Maori or whether to find out more about the mathematics of Maori society and use this as the starting point.

2.4 Holistic Teaching

One aspect of traditional Maori education is that education was not split up into a number of different subject compartments but that a more holistic view was taken. Thus, in designing a mathematics course as part of a general education, some cross-curriculum learning themes may be needed and the mathematics that is appropriate to each of these needs to be developed.

The Maori language syllabus contains fifteen themes which the mathematics curriculum planners believe may be useful. These are genealogy, the school, the body, recreation, the town, food, place of origin, events, bereavement, work, travel, legends, the changing world, aspirations and famous people.

Starting with these it is easy to construct a matrix to find what mathematics could be suitably considered within each theme and what topics need to be dealt with separately. This approach is at present being trialled. Five rows in the matrix are given here as examples.

	Number	Measurement	Algebra	Statistics	Geometry
Recreation	*Number games and puzzles *Scoring systems *Cost of recreation	*Sizes of playing fields *Time, distance & speed		*Games draws *Chance *Odds *Lotto & lotteries	*Board games *Tangrams *Kites *String games
The town	*Money *Shopping *Unit costs	*Mass (weight) *Distance		*Survey of town's facilities	*Trade logos *Map of town - scale factors (or tribal district) *Widths of rivers,etc
Food	*Budgetting (Koha)	*Mass, volume, temperature & time in cooking	*Growth curves for plants & animals *Cooking times, oven & microwave	*Analysis of food *Conservation & sampling	*Packaging food
Work	*Hourly rates, taxes *Percentages *Wages, salaries, commission	*Measurement required in different careers		*Survey of occupations *Graphs of promotion prospects	
Travel	*Budgetting *School trips & camps	*Distance, speed	*Travel graphs		*Maps *Bearings, navigation

3. TEACHING METHODS

Another aspect of mathematics education is concerned with the teaching styles that are used. Some evidence indicates that Maori students do not like to be singled out as better or worse or different from their peers and this suggests that group work should be encouraged to a greater degree than has been done.

Maori students tend not to question authority figures, consequently it is unreasonable to expect these students to initiate a debate with the teacher or sometimes even to ask for assistance.   Other teaching strategies need to be built up so that students are encouraged to discuss and to question amongst themselves and for groups to formulate questions where explanations are needed.

Maori culture was historically an oral culture while most European learning reflects a written culture.   A need exists to find out what a modern Maori feels comfortable with.

4.   RAMIFICATIONS FOR OTHER SUBJECTS

Teachers of other subjects are also considering suitable holistic approaches that incorporate their content and science teachers are considering using the language themes as a starting point.   This means that mathematics teaching could be integrated with Maori language and with science.

If other subject teachers do the same it will become important for the teachers to know what is happening in all the aspects of each theme.   This will put pressure on teachers and may lead to either a need for more general teachers, rather than subject specialists, or for more communication between teachers.

5.   TEACHER SUPPORT

Two networks of teachers have arisen, one a bi-cultural network of over 100 teachers from 80 of the 400 high schools in New Zealand, and the other a subset of it, a bilingual network involving 30 schools.

These networks include both Maori and non-Maori teachers and have developed as a grass-roots movement of teachers who want to see an increase in the bi-cultural and bilingual emphasis in the mathematics curriculum.

To help share the resources that are developed and support these networks two mailings are regularly organised, *Te Kupenga* for the bi-cultural network and *Nga Mauranga* for the bilingual network.

A start has been made, ideas are being tested, resources are being prepared and shared.   The curriculum development division feels that this work is an important priority and the support that teachers have given *Te Kupenga* suggests that these first steps are appreciated.

REFERENCES

Knight, G. (1985).   The Geometry of Maori Art − Rafter Patterns, 36-40.   In *The New Zealand Mathematics Magazine*, **21**, **No 2**, New Zealand.   Auckland Mathematical Association Inc.

Rikihana, T. (1985). *Mathematics* (Draft Edition). New Zealand, Department of Education.

Te Kupenga 1-3 (1987-8) by members of Te Kupenga network. New Zealand, Department of Education.

Nga Mauranga 1 (1988) by members of Nga Mauranga network. New Zealand, Department of Education.

Werry, B. and Knight, G. (1984). The Symmetry of Kowhaiwhai. In *Mathematical Digest*, **No 77**, 1-2. New Zealand.

CHAPTER 9

Transformation Geometry Concepts in Children's Spontaneous Pattern Painting in the Primary School

D. Booth
Narrabeen, Australia.

ABSTRACT

Pattern painting is the child's non–figurative art–form or spontaneous design making. It is a logico–mathematical development passing through three stages: scribble, topology and geometric pattern. There are two classes of patterns, based either on a repetition of an element or on a division of the plane. A number of intuitive mathematical concepts are embedded in the patterns and the symmetry operations of translation, reflection, rotation, underlie their production. The paper describes what pattern painting is and how it develops, and shows the natural link between children's spontaneous pattern painting and transformation geometry. It suggests how the intuitive mathematical concepts embedded in the paintings can be used as a starting point for discussion and the introduction of formal mathematical concepts.

1.   INTRODUCTION

The purpose of this paper is to show the natural link between children's spontaneous pattern painting – an important aspect of child art hitherto largely ignored by educators – and transformation geometry. The paper is based on many years of original observations in pre–schools, on subsequent studies in infant–schools (Booth 1975, 1980) and on experimental research (Booth 1981, 1984, 1985, Booth et al 1988, Smith and Booth 1988). It is an activity suitable for all primary grades from kindergarten to year 6, and even beyond. Pattern making, or geometric decoration, is an activity that exists in every indigenous culture, past and present, throughout the world. It is evident in body painting, wood carving (for example in the decoration of weapons, hunting tools, totem poles), painting of Easter eggs, decorations of ancient pottery, weaving, embroidery, and so on, in both Eastern and Western cultures. It is a reality of universal human activity.

Underlying most geometric decoration are the symmetry transformations. The focus of this paper is to describe briefly what pattern painting is, how it develops, and how it can be used as an interesting teaching aid both in art and geometry in the context of the classroom.

2. WHAT IS SPONTANEOUS PATTERN PAINTING?

Spontaneous pattern making is an aspect of children's play that can be observed in diverse activities such as block play, printing, collage, drawing, painting, and other situations. My research has focused on children's spontaneous pattern creations made with paint, hence pattern painting.

Pattern painting is the child's non-representational or non-figurative art-form and in essence is spontaneous 'design' making. This creative art-form arises completely spontaneously if young children are not influenced by peers or adults to paint figurative pictures. It is a constructive process arising entirely from a child's interaction with the painting materials. Basically, pattern painting is a developmental logico-mathematical activity.

2.1 Development in pattern painting

Pattern painting exhibits three qualitatively distinct and developmentally invariant stages, which I have designated:
1. scribble,
2. topology, and
3. geometric pattern, or pattern for short,
(Booth 1975, 1980, 1981, 1982). Recently Crawford (1988) has observed similar learning stages among pre-schoolers using a LOGO computer programme.

2.1.1 *Stage 1: Scribble*

In this stage the colours are piled more or less in the middle of the paper, leaving the remaining area unpainted. The brush is moved in any direction up and down and from side to side often producing oscillating lines (see figure a). Essentially this stage is a manifestation of developing motor skills necessary for handling the painting materials. Primarily it involves adaptation to the brush and acquisition and coordination of several skills, eg loading the bristles with paint or washing the brush before loading it with another colour. In this stage the child simply concentrates on making marks. When the actions are coordinated and the scribble stroke is established, attention from mark making shifts to *colour* and *space*. It is this shift of attention that ushers in the topology stage.

*FIGURES: (a) scribble, (b) topology, (c) topology

Class 1 Patterns: (d to j) arising from a systematic repetition of an element:

    (d) translation, (e) diminution, (f) and (g) 2D translation (h) 1-fold reflection, (i) 2- or 4-fold reflection or rotation, (j) rotation.

Class 2 Patterns: (k to r) arising from a division of the plane:

    (k) no repetition, (l) 1-fold reflection, (m) and (n) 2- or 4-fold reflection or rotation, (o) not repetition, (p) rotation, (q) 1-, 2- or 4-fold reflection or rotation, (r) rotation.

*Figures are based on original paintings

2.1.2 Stage 2: Topology

'Topology' here is defined as relating to the children's exploration in 2D space. This stage differs from the scribble in three important aspects. First, colours are separated. Second, the whole of the paper surface is covered with irregular shapes placed in irregular order (figure b). Third an attempt is made not to overlap boundaries of adjoining colours. Sometimes patches are free-floating with dots painted around the boundaries (figure c). Sometimes groups of dots are painted within boundaries. These actions are concentrated on and practiced and gradually coordinated, forming a topology painting scheme. Topological notions of 'neighbourhood', 'boundaries', 'inside', 'outside' and 'closed curves' underlie this stage.

In the beginning of this stage the painting is started more or less in the centre of the paper. Later, and particularly towards the end of this stage, the painting is often started close to an edge. It is along the edge of the paper that the scribble stroke becomes modified, or accommodated in the Piagetian sense (Piaget 1953), and becomes visible as a line. Sometimes this shows as a border following the edge of the paper (figure c) or shows as two or three lines along one edge of a topology painting (figure b). In time, the oscillating scribble strokes that have position and direction in space, give rise to lines, dots, and geometric figures. These become elements in the pattern making stage.

2.1.3 Stage 3: Geometric pattern

This stage differs from the preceding one in that the surface of the paper is covered with *regular* shapes placed in *regular* order. A variety of patterns may arise which fall into two classes (Booth 1981):
1. patterns arising from a systematic repetition of an element (examples are given in figures d to j),
2. patterns arising from a systematic division of the plane (figures k to r).

Patterns are built up by a process of matching shapes and are physically generated by symmetry operations or transformations. These operations develop in the order of translation, reflection and rotation. The first simple patterns to arise are translation patterns consisting of lines (figure d) or dots. These might be followed by diminution – so called because children first follow the edge of the paper and systematically work towards the centre (figure e), or translation in 2D (figures f and g), or be followed by 1–fold reflection (figures h and l). Depending on the use of colour, some patterns are 2– or 4–fold reflections and can be rotations also (figures i, m, n and q). Rotations shown in figures j, p and r seldom arise before year 3 when children are about 8 to 9 years of age.

3. METHOD
In order for pattern painting skills to develop, and mathematical and

artistic learning to take place, consistent practice at least once a week is essential. Organising painting lessons is easy if materials and equipment are readily available and pupils are trained to set up their equipment and look after it themselves (for more details see Booth 1987). Each pupil needs: A personal box for paints of *red, yellow, blue* and *white*, a brush, a small water pot, art paper, a palette and a sheet of newspaper to cover the table.

3.1 Organising the lessons
No subject matter is given. The emphasis is on *exploration* and *discovery*. Teaching is principally on an individual level. The teacher moves among the students encouraging any experimentation with mixing colours and creating textures and shapes and praising the students efforts. During and after painting students are asked to reflect, similarly to Southwell (1988), on what they did to achieve certain tones or secondary colours, or textural effects, and to discuss their discoveries in their own words. Gradually the teacher introduces mathematics and art language as detailed in section 3.2.

Class lectures are often confined mainly to procedures and management of materials, but are also used for students to discuss their own paintings with the class. Attitudes to all students must remain encouraging at all times so that no inhibitions develop about ability to paint. Such inhibitions could easily transfer to the learning of the mathematical concepts one is attempting to teach.

3.2 Developing the mathematical concepts in the classroom
From the beginning, pattern painting is a spatial learning experience. It is the exploration of a two-dimensional space to be covered with paint. A number of *intuitive* mathematical notions emerge during the developmental progression. These include notions of classification, sets, enumeration into groups, visual estimation of size (measurement), division of the plane, some Euclidean concepts, and symmetry transformations.

Intuitive topological and classification notions begin to emerge in the topology stage. Topological attributes are simple qualitative relations like proximity and separation, boundaries and regions, open and closed curves, which remain invariant under transformation (Hart and Moore 1973). Topology-stage paintings exhibit these relations. In the paintings, children in effect are doing topological transformation of one irregular shaped area into another. This stage is also the beginning of developing notions of *set* and of *classification by colour*. Classification shows in the child's focus on colour by painting separate colour patches; notions of set show when contrasting colour sets of spots are painted inside the boundaries of patches.

In the pattern stage the irregular shapes are transformed into regular shapes, and the irregular order is transformed into a regular order, but the underlying topological relations remain invariant. Also underlying this stage are intuitive Euclidean notions of congruent shapes, parallel lines, equal angles, equal distances. And implicated are developing skills in visual estimation of size, or measuring by eye. Other intuitive geometric notions include vertical, horizontal, and diagonal lines, and a number of figures, for example, spirals, triangles, rectangles and squares. Cognitive processes of classification now encompass colour, shape and texture, and patterns consist of sets of shapes in different colours. Also, as noted earlier, symmetry operations underlie the production of geometric pattern making. They also underlie the child's aesthetic development (Booth 1976).

The pattern paintings thus exhibit a rich source of intuitive mathematical and art concepts which can be used as a starting point for discussion. For all age levels it is best to use the discoveries in mixing colours as the starting point in teaching, bringing in concepts of primary and secondary colours, followed by concepts of lines (vertical, horizontal, diagonal, parallel, meandering and so on) and geometric figures or shapes, concepts of congruent and similar and finally concepts of symmetry and transformation. At first pupils should be encouraged to use their own informal language to describe their work to the class. Soon after, the teacher may introduce formal language, gradually building up a mathematical vocabulary. Lists of art and mathematics words, both informal and formal, should be compiled during discussion and displayed for reference at all times. Also, after having finished a painting, pupils can write about their painting; what new discoveries they have made about mixing colours, about different ways using the brush to create interesting textures, about what figures and symmetries they have used, the direction and type of lines or figures, or any other mathematical configuration or concept the teacher can notice embedded in a painting. In these ways students become actively engaged – through their own pattern paintings – in manipulating, in reflecting and communicating, and in reading and thinking about mathematical concepts.

4. CONCLUSION

Pattern painting has been found to be an exellent medium for integrating mathematics with other subjects, and it is an excellent way to teach mathematics in a real context – the context of children's own creativity.

REFERENCES

Booth, D. (1975). Pattern Painting by the Young Child: A cognitive developmental approach. Thesis (MEd), University of Sydney.

Booth, D. (1976). Pattern Painting by the Young Child: The roots of aesthetic development. *The Australian Journal of Education*, **20(1)**, 110–112.

Booth, D. (1980). The Young Child's Spontaneous Pattern Painting. In Millicent Poole *Creativity Across the Curriculum*. Sydney: George Allen and Unwin, Australia.

Booth, D. (1981). Aspects of Logico-Mathematical Thinking and Symmetry in Young Children's Spontaneous Pattern Painting. Thesis (PhD) La Trobe University.

Booth, D. (1982). Developmental Sequence of Symmetry Structures in Patterns Painted by Young Children. In Toni Cross and Lorraine Riach (eds) *Issues and Research in Child Development*. Melbourne: Institute of Early Childhood Development, MCAE.

Booth, D. (1984). An Experimental Study on Pattern Painting by Kindergarten Children. *Journal of the Insitute of Art Education*, **8(3)**, 19–24.

Booth, D. (1985). Art and Geometry Learning Through Spontaneous Pattern Painting. *Journal of the Institute of Art Education*, **9(2)**, 38–42.

Booth, D. (1987). Children's Pattern Painting. *Art in Education* (Journal of the Art Education Society, NSW), **12** May, 20–31.

Booth, D., Conroy, J. and Merryman, J. (1988). Learning Transformation Geometry through Spontaneous Pattern Painting. Report to the Management Information Services, Development of Education, New South Wales Government.

Crawford, K. (1988). New Context for Learning in Mathematics. *Proceedings for the Twelfth International Conference for the Psychology of Mathematics Education*, **V1**, 239–246. Veszprem, Hungary, 20–25 July.

Hart, R. A. and Moore, G. G. (1973). The Development of Spatial Cognition. In Downs, R. M. and Stea, D. (eds) *Image and Environment*. Chigago: Aldine.

Piaget, J. (1953). *The Origin of Intelligence in the Child*. 2nd edn. London: Routledge and Kegan Paul.

Smith, I. and Booth, D. (1988). The Effect of Two Approaches to Art Curriculum on Student Self-Concept. Paper presented to the *1988 AARE National Conference*, University of New England, Armidale, 30 November – 4 December.

Southwell, B. (1988).   Construction and Reconstruction: The reflective practice in mathematics education.   *Proceedings of the Twelfth International Conference for the Psychology of Mathematics Education,* **V2,** 584-586.   Veszprem, Hungary, 20-25 July.

CHAPTER 10

Analysing Data

H. W. Henn
Lessing-Gymnasium Karlsruhe, FR Germany

ABSTRACT
Evaluation, analysis and interpretation of data is an important task of all sciences. Students should be introduced to a critical relationship with numbers. As an example, the two aspects measurement of a single value and investigation of the functional interrelationship of two measured values are discussed. The current mathematical subject should be applied, practised and constantly repeated. The proposed examples are interesting experiments, which allow us to mix empirical and mathematical conclusions.

1. AIMS OF THE PAPER

In all sciences and their technical applications, a very important task is not only to collect but analyse and interpret data (in a broad sense). Using pocket calculators and computers, high school students (and scientists and politicians) tend to compute, uncritically, numbers with high precision and accept these as correct and reasonable or to extrapolate, by poor judgement, a few values obtained by measurement to an expected functional relationship. Students should be taught from the beginning to handle numbers critically (Henn (1988)).

As an example I would like to discuss the following two aspects.

(a) A quantity is to be obtained by measuring it directly, or by computing it from measured quantities. How to get the best possible result from the measured data?

(b) How to conclude from the data (x_1/y_1), obtained by measurement, which functional relationship applies between the two quantities?

The mathematical arguments for solving the problems proposed and the examples, are designed and tested for lower secondary level. Since the stress is on processing and interpreting the set of measured data, the discussion should be completed in a maths–class. Of course, analysis of data must not be an additional subject, but should be developed only as an application and repetition of the current subject. Pocket calculators and computers are necessary to process extensive lists of numbers, but this is not reasonable without a clear understanding of the 'real' numbers found in reality (as opposed to the 'ideal' numbers of mathematics). Mixing empirical and mathematical conclusions, a procedure is developed which is significant not only for natural sciences but also more and more for social sciences and humanities.

2. MEASURING A SINGLE QUANTITY

2.1 Specification of exactness

An important fact for students (which is far from evident for them) is that the measured or computed result is not the true quantity X but only a more or less precise approximation. Not only students miss this point as the following example shows: the US–mail service reports the size of a newly issued stamp to be 48.768×43.434 mm^2, thus it is precise up to μm. This meaningless precision can probably be explained by an uncritical computer conversion of the original 1.92×1.71 square inches to the metric system. Even at an early stage, the customary notation of natural sciences should be applied: the $=$ of physics, such as $s = 3.47$ m is mathematically only $s \approx 3.47$ m, meaning 3.465 m $\leqslant s < 3.475$ m, that is, not a point but only an interval of the real line is described. Therefore we must not confuse 'ideal' numbers like $2 = 2.0 = 2.00$, $\sqrt{2}$, π and 'real' numbers $2 \neq 2.0 \neq 2.00$, 1.4, 3.14 saying different things about the measurements. Handling rounded numbers and valid digits intelligently is difficult for students as experience shows (Fanghänel et al 1977).

The precision of a single measurement is restricted by the instrument used, eg mm using a yardstick. If a result has been computed from multiple data inputs by a pocket calculator, the result should not be calculated to more valid digits than the least precise single measurement has. A better method (easily practicable with the help of a pocket calculator) is to compute three results: in addition to the usual computation, a lower bound, using those bounds of the single data giving the least result, then the highest possible result. Now one can give a reasonable mean value with bounds for the error. Students are very often surprised how large this interval of the possible error may become.

2.2 Mean Value

To find a quantity X more reliably, one measures it repeatedly and takes the arithmetical mean $\bar{x}$ as a plausible value of it. To justify this one has to talk about the difference between random and systematical (eg a too short yardstick) errors of measurement.   To get a measure for the precision of the set of data x_1, x_2, ..., x_n one has to apply parameters for the deviation.   As usual (Schupp et al 1980) one arrives at the standard square deviation $(x_n-\bar{x})^2$ as a parameter which is easily treated mathematically and stresses large errors more than small ones.   Already in the lower secondary level one can give a satisfying mathematical reason for using the arithmetical mean by showing that it is the estimate minimising $Q(x) = \Sigma(x_n-x)^2$.   Looking at the variance $s^2 = \Sigma(x_n-\bar{x})^2/n$ and, with respect to the unit of measurement, the standard deviation $\sigma = \sqrt{(s^2)}$ is the mean deviation of the data from the arithmetical mean, and thus gives the measure of precision.

To quantify how much the mean is more accurate than a single value one can introduce the standard error $\Delta x = \sigma\bar{x} = \sigma/(\sqrt{n})$ by mathematically elementary, but not so easily grasped, quantitative consideration.   The result, then, is written $\bar{x} \pm \Delta x$.

2.3 Example: Outdoor measurements

Outdoor measurements in connection with the intercept theorems (in Germany in class 9) or with trigonometry (class 10) are a very apt theme to reach the aims discussed here.   Students find the height of the school building using the "Jacobsstab", one of the oldest instruments used for measuring angles at sea.   The class measured the height of the school building up to the highest window (so that we could measure the distance afterwards directly) with different, self-made instruments.   We have, by the intercept theorems, $h = (1 + b/a)\cdot s$.

The results of the individual students for h differed considerably from 12.0m to 16.3m.   To estimate the error of a single measurement, eg $h = 14.3$m, we estimated the bounds of the error in a, b and s:

distance of nails on Jacobsstab: $s = 10$cm $\pm$ 0.2cm
distance a on Jacobsstab: $a = 18$cm $\pm$ 1cm
distance b Jacobsstab
                    – school building: $b = 25.6$m $\pm$ 0.3m.

Entering the mean value of s, a and b in the formula for h and those bounds giving a maximal h_{max} and a minimal h_{min} one gets

$h = 14.31$m, $h_{max} = 15.64$m, $h_{min} = 13.15$m.

The single measurement is therefore rather inaccurate; as a result one can give at most $h = 14.3$m $\pm$ 1.3m.

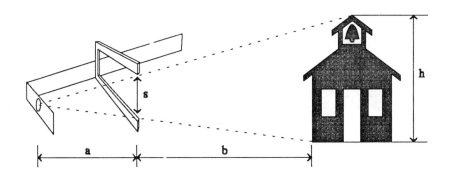

Figure 1

Students had measured the height independently. From these data we computed the mean $\overline{h}$ = 14.75m and the standard error Δh = 0.27m. The improvement is obvious! Finally it was interesting to compare the correct value h = 14.45m found by direct measurement holding a yardstick out of the window.

3.   TESTING A LINEAR RELATION BETWEEN TWO QUANTITIES

3.1 Graphical method
Two variable dependent quantities X, Y are measured. The data obtained (x_1/y_1), (x_2/y_2), ..., (x_n/y_n) are entered in a diagram of appropriate scale. A linear relation can be tested graphically using a rule, and a mean line can be drawn. Sometimes it is clear from the context that the line meets the origin. The accuracy of the measurement and the exactness of the linearity may be judged graphically from the points and line drawn. It is most important to teach this method of analysing the data in the classroom. The calculation of the mean line exceeds the mathematical and temporal possibilities of the lower secondary level.

Students tend to connect the points drawn in the diagram by a polygon. For them the points are something absolute. It is a considerable intellectual achievement to accept the linear relation as a hypothesis and thus draw a mean line. The points drawn become a more or less exact approximation (containing errors of measurement) of the "true" points.

3.2 Linearisation

If the points in the diagram are positioned such that a linear hypothesis obviously does not make sense, a linearisation must be tried. This means that another functional relationship (inspired by a curve approximating the points) is chosen as a hypothesis to be tested. The function should not be chosen at will, but must be adapted to the situation in question. A physicist will accept a formula deduced from a mathematical model if the measured points fit it satisfactorily but will reject a function for which he cannot give a reason, even if it approximates the points slightly better.

Then one tries to obtain a linear relation $w = mv + b$ by a transformation of the variables x,y into new variables v,w which is as simple as possible. Often it is clear, by some reasons peculiar to the problem, that the line will pass the origin.

If a straight line now passes (approximately) through the points (v_1/w_1), ..., (v_n/w_n) the hypothetical function has been confirmed, otherwise it has to be rejected or revised.

Finding empirical formulae, as described, can be very difficult, especially for the lower secondary level, and often the teacher will have to tell the class what to do.

3.3 Example 1: The dropping ball

A ball dropped on the ground does not return to the initial height h but only up to a smaller height h' because it loses energy by bouncing on the floor (Henn 1985). If the data (h/h') are drawn as a diagram the points suggest a linear relation between h and h' : $h' = mh$. The graph is a line through (0/0) with gradient $0 \leqslant m \leqslant 1$. The cases $m = 1$ and $m = 0$ describe an ideally elastic or totally inelastic ball. This example can be used very well in class 7 (topic: proportionality) or class 8 (topic: equation of the straight line). The goals of instruction are: how to evaluate measurements, how to find functional relations, and to discuss the mathematical model and its limits. Figure 2 shows the work of a student. The dotted curve is the polygon-like connection of the points. The full time is the mean line.

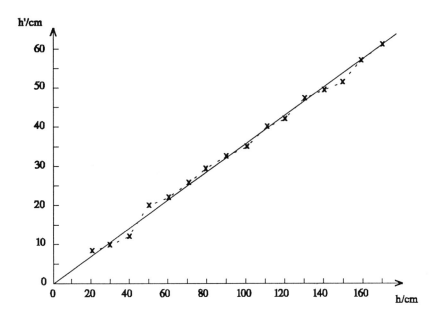

Figure 2

3.4 Example 2: Measurement to determine the specific resistance

It is physically plausible that the electrical resistance R of a wire depends on its length ℓ, its diameter d and its material. Here, one fixes the length ℓ (= 1m), varies d and measures R. One obtains a curve like a parabola (Figure 3). In order to linearise, one first tries $v = 1/d$ but the result in Figure 4 shows that this hypothesis has to be rejected. Only the linearisation $v = 1/d^2$ furnishes points on a straight line. It is clear by physical argument that a line R = mv, $v = 1/d$ or $v = 1/d^2$ passing through (0/0) has to be used: $v = 0$ means infinite diameter for which R = 0 is reasonable. The final result says that the resistance is proportional to the inverse cross section area of the wire (which is plausible but not obvious from the outset nor a general law – catch word "skin effect").

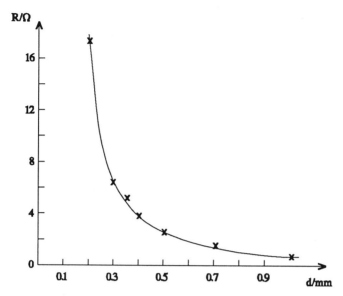

Figure 3

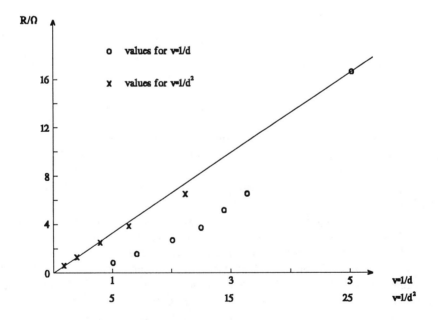

Figure 4

3.5 Example 3: Decay of beer foam

Especially important and interesting are measurements concerning an exponential law (Sennekamp 1980). A suitable experiment to be done at home is the decay of beer foam: the student pours as much beer foam as possible in a tall beer glass and then measures the height of the foam once every twenty seconds. Figure 5 shows data as found by a student. The points suggest the idea of a falling exponential curve. This hypothesis must be verified by a fitting conjectural function and linearisation. An exponential relation between height h and time t may be written $h = H(t) = b \cdot a^t$.

Since H is decreasing there must hold $0 < a < 1$. The constant b can be interpreted as the initial height but does not yet have a physical interpretation. The question when half of the initial height is reached suggests base 2 in the formula but then one has to insert a constant c in the exponent: $h = H(t) = b \cdot 2^{-ct}$.

For $t = 1/c$ the foam just has shrunk to half its original size. Thus the constant c gets a physical interpretation: $T = 1/c$ is the half life. The resulting formula

$$h = H(t) = b \cdot 2^{-t/T}$$

is much more transparent for students. The equation is linearised by taking the logarithm:

$$\log h = \log(b \cdot 2^{-t/T}) = -((\log 2)/T) \cdot t + \log b.$$

We have an exponential decay if, and only if, the quantities (v/w) with $v = t$ and $w = \log h$ are on a straight line. The gradient m of this line gives the half life because $T = -(\log 2)/m$. This linearisation is shown in Figure 6. A linear approximation makes sense only for $0 \leqslant t < 280s$. The half life is $T = 130s$.

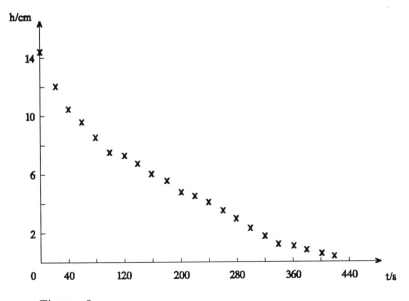

Figure 5

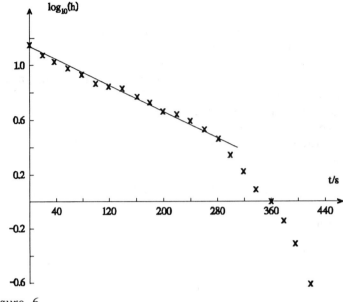

Figure 6

Last but not least (for this example): It is important for the students to understand that mathematically all bases for exponential functions are of the same value but for the applications they signify different things: In physics, if time is the independent variable, basis 2 is most reasonable, but for computing the derived function or integrals basis e is practical (this is why e is preferred as base in mathematics).

REFERENCES

Fanghänel, G., Flade, L. and Pruzina, M. (1987). Nochmals zum Thema "Sinnvolle Genauigkeit". *Mathematik in der Schule*, **25**, 465–477.

Henn, H. W. (1985). Ballspiele, mathematisch und didaktisch betrachtet. *Journal für Mathematikdidaktik*, **6**, 211–226.

Henn, H. W. (1988). Meßwertanalyse – Eine Anwendungsaufgabe im Mathematikunterricht der Sekundarstufe 1. *Der mathematische und naturwissenschaftliche Unterricht*, **41**, 143–151.

Schupp, P., Schweizer, U. and Wagenknecht, N. (1980). Beschreibende Statistik. *MS 1. Tübingen: Deutsches Institut für Fernstudien.*

Sennekamp, B. (1980). Exponentielle Vorgänge: Kaffee filtern und Bierschaumzerfall als physikalische Schulversuche. *Der mathematische und naturwissenschaftliche Unterricht*, **33**, 31–36.

CHAPTER 11

Mathematics of a Lake – Problem Solving in the Real World

T. Matsumiya
Department of Mathematics, Osaka Kyoiku University, Japan
A. Yanagimoto
Tennoji J H School, Affiliated to Osaka Kyoiku University, Japan
Y. Mori
Tennoji J H School, Affiliated to Osaka Kyoiku University, Japan

1. PROBLEMS

The problems to be solved in mathematics education are summarised as follows.

(a) <u>The problems of reality</u>
Since the modernisation of mathematics education, the contents dealt with at school have become more and more abstract and imaginary, where school mathematics seems to be treated as one part of the logical, abstract and formal aspects of mathematics. Under such mathematics education, however, it is impossible to teach the slow learners found everywhere today. Therefore, it seems to be necessary for us to change the view. Based upon the real world around the learners, we should try to set up new mathematics which would enable learners to apply mathematics to real world problems. What is necessary is not only dealing with theoretical mathematics but giving the learners opportunities to solve problems in the real world through mathematical ways of thinking.

(b) <u>The problem of actual practice</u>
As described above, school mathematics has been an abstract theory. It is also important to teach learners how to conduct experiments, hand manipulation, and how to manipulate teaching tools (including

using personal computers) even in junior high school. Such new school mathematics works not only as a device for the learners to understand mathematics, but also contains some essential aspects of the mathematisation of things and affairs in the real world and the application of mathematics to them.

(c)   The problems of totality

In the mathematics classroom of today, the students usually learn numbers, algebraic expressions, functions, probability and statistics, geometrical figures and so on, one after another, separately. They are given almost no opportunity to learn them totally or use them together to solve a problem. That is to say, it is just a learning of each item, which increases the number of learners who do not understand the reasons why they should study mathematics. Actually the number of such learners increases more and more as their grade proceeds.

In general, has our education been teaching mathematics just as knowledge or techniques? What should we do to change our teaching direction? This has been our central subject since 1982. This is why we chose the theme of this present paper.

2.   VIEWS TO SOLVE

The materials used in the classroom should fulfil the following conditions:

(a)   realistic materials,

(b)   materials that contain hand manipulations (experiments, drawing figures, making solids, measurements, etc) or using teaching tools (including personal computers),

(c)   materials to which the learners have to apply what they have learned totally,

(d)   materials that can be finished in a few hours.

We call learning with these materials the Composite Real Mathematics approach (hereafter we use CRM). We intend to teach with them twice or three times a year. This is an attempt to get over the problem solving, in which solving itself is the crucial part. That is, this is an attempt to use the mathematical scientific approach, and to teach mathematics as a humanistic education.

1st grade 12-13 years old	2nd grade 13-14 years old	3rd grade 14-15 years old
Integers	Algebraic Expressions	Algebraic Expressions
Positive and Negative Numbers	Linear Inequations	Square Root
Algebraic Expressions	Simultaneous Equations	Quadratic Equations
Equations (ax+b=c)	Linear Functions	Quadratic Functions $(y=ax^2)$
Functions $(y=ax,\ y=a/x)$		
CRM_1	CRM_3	CRM_5
Plane figures	Congruence	Circle and Angle of Circumference
Solid geometrical figures	Similarity	Pythagorean Theorem
Measurement (length, area and volume)	Statistics (mean value and range)	Measurement (application of similarity)
		Probability and Sampling
CRM_2	CRM_4	CRM_6

Table 1:   Curriculum of Mathematics for the Lower Secondary School
(Japan 1981)

Table 1 is a curriculum of mathematics for the lower secondary school in Japan in which we set six CRM projects.   In the 1st grade, we think that we may practise CRM_1 and CRM_2 if we find good materials.   We need not always practice them in the 1st grade.   We planned Mathematics of a Lake as CRM_6 in the 3rd grade.

3.   SUBJECTS AND TEACHING PLAN

3.1 Subjects

Restriction of the water supply from the Yodo River (which comes from Lake Biwa, the largest lake in Japan) to Osaka City has been imposed five times since 1973 because of water shortage.   Since the river supplies water to 11 million people and many factories in the Kinki District, it is valuable to learn the problems related to the lake which is a tank supplying water to the people and the factories.   Therefore, we attempted to construct the CRM, making this lake as a subject according to the views above.

3.2 Teaching Plan

Table 2 is an example of our actual teaching that we intended to practice to place at the end of the 3rd grade (age 14 and 15).

Unit		Allotment of Time	Specific mathematics
1st	Volume of water in the lake and area and volume	3 periods	Statistics and measurement
2nd	Changes of water level and linear function	3 periods	Linear function
3rd	Circumference of the lake and length of curve	2 periods	Circle and measurement
4th	Volume of used water and statistics	2 periods	Statistics

Table 2:   Teaching Plan of Mathematics of a Lake

Personal computers and calculators were used effectively in each unit.

We introduced the students to Mathematics of a Lake as follows. Some students had experienced the restriction of water supply from the Yodo River, and many students knew the phenomenon of the water shortage in Lake Biwa.   So, we suggested that they should think about this theme.   We, the teacher and the students, then discussed it.   The water shortage in the Yodo River which comes from Lake Biwa has, of course, a direct relation to that of the lake.   It is the key to the problem of the water shortage, whether we can predict and control the water level of the lake.   So we started to think about the quantity of water in the lake.   What is the water volume of Lake Biwa?   How is the water level of Lake Biwa changing?   How is the drainage from the gate between the lake and the river controlled?   When the students were solving these problems we suggested that they used the mathematics which they had already learned, and, if necessary, we taught them something new.   We then gave some problems which appear in section 4.

4.   EXAMPLES OF PROBLEM SOLVING ACTIVITIES

4.1   The content of the 1st unit [volume of water in the lake, area and volume]

Problem
The problem was to find the area and volume of Lake Biwa.

Solution method
We discussed with students how to solve this problem.   As a result, we found out the following method.   Then the students did the solving activities.   They used the following specific mathematics contents: Monte Carlo simulation and the volume of columns.

Reference materials
Figure 1, the map of Lake Biwa, was given to the learners.   A map drawn on a scale of 1 to 250,000 was given as it was just the size of the computer's screen.

Calculation of the area
First the area enclosed within the line of the lake shore was   found.   A personal computer was introduced as it was very difficult to calculate the indefinite-shaped area.
      The line was copied on the screen from the map drawn on a scale of 1 to 250,000.   Then the area of the figure on the screen was calculated using the personal computer.   That is, the personal computer put dots randomly on the screen and estimated the ratio of the number of dots within the figure (Monte Carlo simulation) to the total area.
      The area that the students found in this way was about 676 km^2. This is approximately accurate.   The real area of Lake Biwa is 673 km^2 according to a geographical datum.

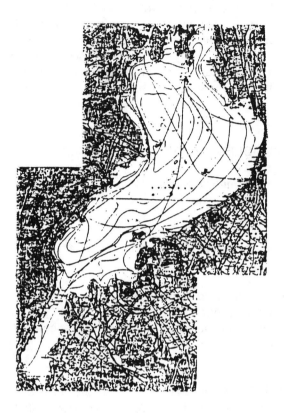

Figure 1:   Map of Lake Biwa and its contour lines

Calculation of the volume
Likewise, each student took one of the figures enclosed by the contour lines (which means, in this case, a line linking the points of the same depth) drawn at the intervals of every 20 metres.   The students were told to regard the volume of the lake as the aggregate of the columns whose area was a figure enclosed with a contour line and whose height was 20 metres.

Thinking of the columns which could be just inside the lake as seen in figure 2, they calculated the volume of these columns.   Consequently the sum of them was about 22.3 km^3.   On the other hand, they also thought of the columns which were just outside the lake.   In this case, the sum of the volume of the columns was about 35.8 km^3.   The average of these two volumes can be found as follows: (22.3+35.8)/2 = 29.1.   According to official data, the water volume of Lake Biwa is 27.5 km^3.   Here the result brought out by the students (ie 29.1 km^3) was also approximately accurate.

depth

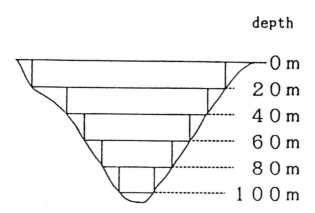

Figure 2:   A longitudinal section of Lake Biwa

4.2 The content of the 2nd unit [changes of water level and linear function]

Problem

The problem was to find how the water level of Lake Biwa was changing. We focussed especially on the following two tasks. One was to find the relation between the drainage from the gate and the water levels of Lake Biwa. The other was to find out the linear functional relations in the decreasing and increasing changes of the lake's water level.

Solution method

We discussed with the students the different ways to solve this problem. We made them aware that we could simulate the changing water level of Lake Biwa by the amount of drainage from the gate and by the past changes of Lake Biwa's water level (figure 3). They used the following specific mathematics: calculation of the quantity of water and linear functions.

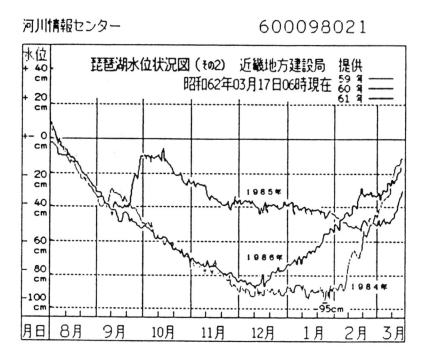

Figure 3:   Changes of the water of Lake Biwa

Reference materials
Figure 3 is the reference material.   The graph shows the changes of the
water level from 1984 to 1986.   The changes in 1984 and 1986 show
abnormal curves that mean water shortage phenomena were observed.
The changes in 1985, on the other hand, show a normal curve.   When
the water shortage occurred, as it did in 1984 and 1986, both the
change of water decrease from August to November and the change of
water increase from February to March seem to show, approximately, a
similar change to that of a linear function.   It is possible, therefore, to
regard the changes as straight lines, which enables us to study these
changes of water level as one of the relations in linear function.

The drainage from the gate and the water levels
The first task was to find the relation between the drainage from the
gate and the water level of the lake.   The task was to find out what
relation they have, by regarding the lake as a column whose side was
vertical to its base and without considering any other elements except the
drainage.   The minimum drainage off the lake is 5 tons per second, and
the maximum is 650 tons per second.   The students found that we could
control the number of days to reach the dangerous level since we could
change the rate of water decrease (ie slope of the graph) by controlling
the drainage from the gate (figure 4).

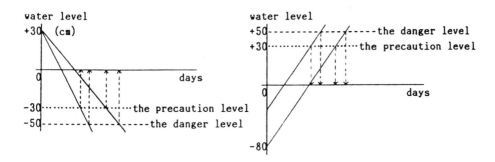

Figure 4   Figure 5

The changes of the water level and the linear functions
The second task was to find out the linear functional relations in the
decreasing and increasing changes of the lake's water level when the
shortage of water was observed.   The students found that they could
predict the number of days to reach the dangerous water level from the
first volume of the water (ie y-intercept of the graph) if the ratio of
the water decrease was fixed (figure 5).
    The students did both tasks above with the help of the computer
simulations.

4.3 The content of the 3rd unit [circumference of the lake and length of the curves]

Problem
The problem was to examine the shape of the lake and to find the
length of its circumference from the results of the examination.

Solution method
We motivated the students to do the task by pointing out that this made
clear the lake's change of shape in the shortage period.   In order to
solve this problem we had them use the following specific mathematics
contents: the nature of a circle (ie idea of curvature, drawing the circle
determined with three points and finding the arc length) and
measurement.

Reference materials
The same map as Figure 1 was given to each student.

Approximation of the shape of Lake Biwa with lines and circles
First, we had them find out how to approximate the shape of an
indeterminate figure.   Then, they found that the circumference of an

indeterminate figure consists of straight lines and curved lines. They also understood that the shape of a curved line could be approximated with a circle and that using its radius was the best way to characterise its curvature numerically.

Next, we made them divide the circumference of Lake Biwa into linear parts and curved parts and draw approximate circles of each curve using a ruler and a pair of compasses. How to approximate them was, of course, up to each student. They plotted any three points on the curve, drew two perpendicular bisectors of the segment between two of them, and drew a circle of curvature around the intersection point. (They had learnt the way to draw the circle determined with three points when they were in the 1st grade). At the same time, we made them measure the central angle of the sector formed by those three points and the centre point. Figure 6 is an example of the approximations drawn by the students. It shows that the radius of curvature in the neighbourhood of B is 1.3 cm and in the neighbourhood of D it is 2.5 cm. In this way they approximated all parts of Lake Biwa one after the other.

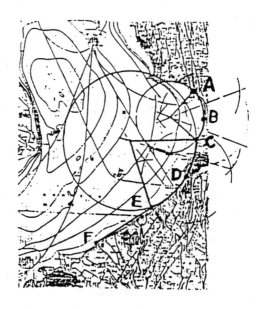

Figure 6

Finding the length of the lake's circumference
Finally, we asked them to find the length of the circumference of Lake Biwa. They soon became aware that it was good to use each radius (r) and central angle (a). Then, by using a calculator they calculated the

arc length $\ell = \pi ra/180$ of the sector at each curved part and found the length of its circumference with the following formula:

$$\begin{bmatrix} \text{total sum of} \\ \text{curved parts} \end{bmatrix} + \begin{bmatrix} \text{total sum of} \\ \text{linear parts} \end{bmatrix} \text{ scale of map}$$

They found that the length was between 200 km and 240 km (the real value is 240 km). They understood the error, caused by the approximations of their work, and they were quite satisfied with their results.

In this unit the students used only calculators, but it is possible to use programmable calculators or personal computers.

4.4 The content of the 4th unit [volume of used water and statistics]
They tried to find the trend of the volume of water used at home through statistical methods. The details of their activities are omitted here.

5. CONCLUSION
The problem–solving activities used in schools of today are rather unrealistic and element–enumerative. This means that too much emphasis is laid upon solving activities themselves without examining problems as well. In order to overcome this limitation in problem solving, we have used the Composite Real Mathematics approach with real world problems as we did in this practice. Through this kind of CRM we intended to solve the problems of reality, actual practice and totality, as refered to above. The results of our practice with the 3rd grade students are summarized as follows.

1. The learners found out a new aspect of school mathematics. Now they understand that it has a close relation with the actual world and it is helpful to their daily lives.

2. Their motivation has been strongly stimulated. Our practice has made them more interested in mathematics.

Finally, we aimed at a drastic change from mathematics education just giving the learners knowledge and techniques to the humanistic education. The objective to be pursued now is to integrate the contents of such learning into mathematical sciences.

CHAPTER 12

From Shadow to Light
An introduction to space geometry at senior school level

B. Parzysz
Université Paris-7, France

ABSTRACT
Studying space geometry at senior school level requires the help of drawings, but making these drawings requires some knowledge of space geometry; this is a perfect example of a vicious circle. In order to break it, and thus to improve the students' understanding of space geometry, we have designed and carried out a didactical research, whose basic idea is as follows.

The shadow cast on a plane either by a light bulb or by the sun, is a suitable realisation of a projection, central or parallel, and its properties are far less well known than one would think.

To study them we use materials allowing us not only to realise the phenomenon but to simulate it. This enables the students to learn something about both shadows and space geometry.

1. INTRODUCTION

(a) Space geometry is supposed to be a difficult subject among senior school students. Within our present research, we could see that part of these difficulties come from the drawings which are used, both as a visualisation of space situations and as a tool for solving problems. This convinced us that improving their understanding of how to read and write space geometry drawings would lead them to a better understanding of space geometry itself. Hence the *need for an explicit teaching of the rules of drawing*, as part of the teaching of space geometry.

(b)   Whenever and wherever possible, the teaching of mathematics should
      be anchored in a physical reality (which certainly is a motivation
      for most students), theorising it in order to solve problems presented
      by that reality.   By a constant dialectic between the theoretical
      model and its realisation, knowledge, both of the surrounding world
      and of mathematics, would be increased.   However, for that it is
      necessary for the everyday problems which they take an interest in
      to be *real* problems, whose solution is not evident and requires the
      use, and perhaps even the making, of mathematical *tools*, which
      later will be able themselves to become *subjects* of study, before
      being reused as tools to solve new problems (the tool/subject
      dialectic).   Geometry at senior school level, and especially space
      geometry, seems favourable ground for putting this into practice.

2.   REPRESENTATION IN SPACE GEOMETRY

2.1 Conventions or properties?
A problem which currently occupies several French research teams in
didactics is that of the plane representation of three-dimensional figures
in the teaching of mathematics.   The classical representation is usually
presented, whenever it is, as resulting from conventions.   However, this
is not the case, and in fact it is based on the properties of projection.
Unhappily enough, the study of projection is a problem of space
geometry, whose solution requires the use of plane representations – a
perfect example of a vicious circle which seems difficult to break.

2.2 Why use shadows and materials?
We think that studying shadows might well be a means of overcoming
this difficulty, for two main reasons.

1.   The shadow cast by the sun allows a good physical realisation of
      the parallel projection on a plane.   As the parallelism of the sun
      rays falling down on a terrestrial object is not always obvious to
      high school students, it seems more interesting to begin with the
      shadow cast by an electric light bulb, which is a realisation of a
      central projection.

2.   This phenomenon, which we come across every day, is so common
      that we may think it holds no more secrets, and hence no more
      interest, for us.   Is this really the case, though?   It is by no means
      certain.   If problems arise where everything was though to be
      known, there is perhaps a motivation for trying to solve them.

Thus the question is asked how students

(i)   can be shown that some points about shadow they believed
             well known still remain (if I may say so) in the dark,

(ii)   can be allowed to cast all the light on the problems
             themselves.

Experience has convinced us that it is impossible, even with senior
school students, completely to do without the use of three-dimensional
objects in the teaching of space geometry so, in order to set the
problem and solve it, we use materials (a model).

3.   LIGHT BULB SHADOW

3.1 Materials and students' conceptions

The materials (figure 1) consist of a board supporting a telescopic post,
on top of which a small electric light bulb is fixed.   On the board lies
a 'skeleton' cube made of wooden sticks, the upper square of which is
painted red.

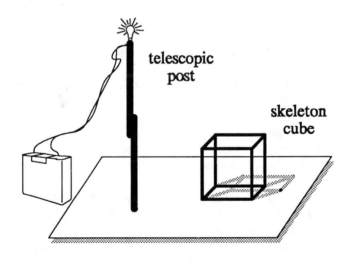

telescopic
post

skeleton
cube

Figure 1

The problem is set to the class in the following form.

If darkness could be obtained in the room, the bulb would cast a
shadow of the cube on the board.   We shall first consider the shadow
of the red square alone.   Please write down your answers to the
following questions.

1. What is the shape of the shadow of the red square when the cube lies on the board in the following positions (figure 2)?

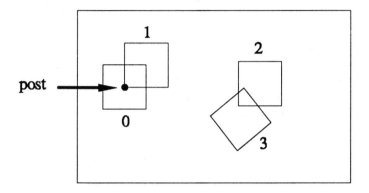

Figure 2

2. The cube is lying in position 2. How does the shape of the shadow vary when:

(a) the bulb is lifted up, without moving the cube,
(b) the cube is moved away from the post, without moving the bulb?

The answers given by the students are very varied, and reflect their naive conceptions about the properties of shadows (theorems and actuality), as for instance generalising incorrectly about the properties of the shadows of vertical objects (figure 3).

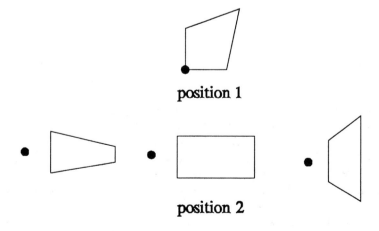

Figure 3

During the debate which takes place after this, the students' different conceptions (most of them being misconceptions) come into conflict, and the divergence of opinions about a phenomenon apparently so common is perceived as a kind of intellectual challenge (destabilisation of knowledge) and arouses the desire to know the truth – what happens *really*?

3.2 Simulation and geometry
Resorting to an effective realisation of the phenomenon being explicitly excluded, on the fallacious pretext that the classroom cannot be totally darkened, we must fall back on a simulation, realised thanks to stretched elastic, joining the bulb to the corners of the red square, the cube lying in postion 3 (the most general case)(figure 4).

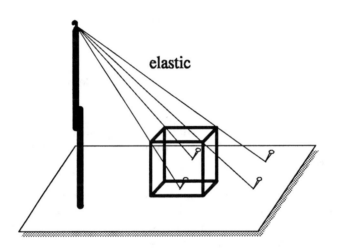

Figure 4

To general astonishment, it seems that the shape of the shadow of the red cube is square (less than one student out of ten gives in fact this answer).   However, to be quite sure, a demonstration would be necessary.

The students think of different methods (consisting of drawings in particular planes made concrete by the elastic), which use spatial properties which are not yet formalised, allow the use of previous knowledge about plane geometry, and lead to the confirmation of the experimental results.   I would point out that with fifth graders we proceed in a more technical way, having them make scale drawings in these particular planes (figure 5).

Indeed this type of activity makes the students more active, allows them to produce some constructions in plane geometry, not 'for pleasure', but because they need them to solve a problem (for instance, a triangle whose lengths of sides are known), and in some way introduces them to a sort of 'technical drawing', in which good co-ordination of the different views – replacing the possibility of looking round the real object – encourages progress in the knowledge of this object.

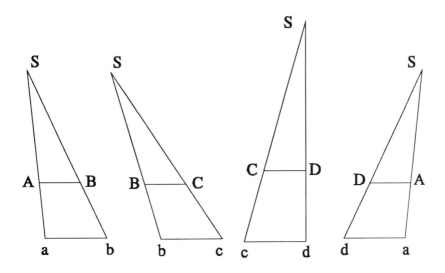

Figure 5

3.3 From shadow to drawing

One is then able to draw the shadow of the cube on the ground, when the position of the light source is defined, and this allows a reinterpretation of the initial answers given by the students to the questions, as well as a study of the properties possibly retained by this type of shadow (mainly alignment, parallelism, middle, ...). The drawings of the shadow look very much like drawings of a cube 'in perspective', as the students call it, and so a link can be established with the painters' perspective (linear perspective), thanks to a comparison of our materials with a 'Dürer's window' focusing on the analogies and differences between the two situations (figure 6).

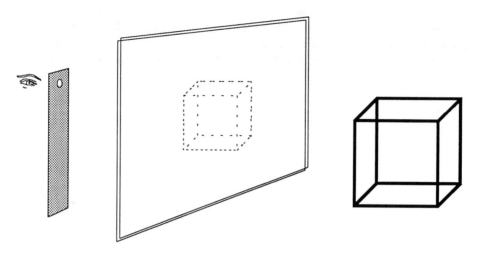

Figure 6

At the same time, we come to the question of 'dotted lines', considering, either the edges of the cube which are seen (window), or the ones which are lit up (model).

4.   SUN SHADOW

4.1 Towards parallelism

Afterwards, the shift towards sun shadow is made by investigating the conditions in which the shadow would retain the middle of a vertical edge of the cube (figure 7).

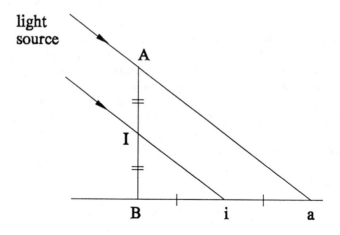

Figure 7

The students find out that this requires the light source to be 'very distant', 'to infinity'; the crucial point, that is parallel projection on a plane, can now be considered.

Even in this simpler case, the students' opinions are still varied, as new questions, very similar to the previous ones, show. The cube is lying on the desk, we ask them the following.

1. What is the shape of the shadow of the red square

   (a)   when it is horizontal,
   (b)   when it is vertical?

2. To make a drawing of the shadow of the cube when the rays of the sun come from particular directions (vertical, diagonal of a vertical side, diagonal of the cube (see figure 8)).

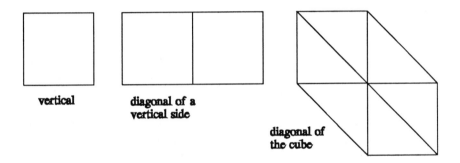

Figure 8

4.2 Simulation and geometry

New materials allow an experimental verification of the answers. They consist of two contiguous boards, one being horizontal, on which the cube is lying, and the other vertical; between these boards, elastic can be stretched, in order to simulate the rays of the sun (figure 9).

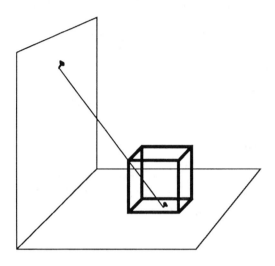

Figure 9

   The problem is now to stretch an elastic parallel to a given elastic and passing through a given point. The different methods recommended by the students enable them to specify the relative positions of two straight lines in space.   Besides, here again the three-dimensional experimental model allows the elaboration of a geometrical demonstration of the results observed and conjectured, with the help of drawings in particular planes. The plane homothety found in the case of the light bulb shadow between the lower side of the cube and the shadow of the upper side becomes here a translation, and so we can come back to the analogy between the two types of transformations, here linked to the projective nature of the situation.   Moreover, we are now able to demonstrate the principal, and so useful, properties of this type of projection: retaining parallelism, centre of gravity, 'real size' image in the planes parallel to the projection plane.

4.3 Parallel perspective
The students thus become aware of the advantages of this kind of drawn representation:

1.   it retains an acceptable visual resemblance to the object, close enough to a photographic image (in other words what is *perceived* of the object);

2.   it retains as well a certain number of properties of the object (in other words what is *known* of the object), properties one is always reluctant to lose and which, besides, make the representation easier

to draw.

Sun shadow provides an interpretation of the 'dotted line' convention as well; although it becomes actually difficult to speak of a point of view in the case of parallel projection, one may still consider without difficulty the parts of the object which are lit up.

Once in possession of the rules of drawing in parallel perspective, the students are in a position to apply them to the traditional drawings which are used in the teaching of space geometry, and to answer questions such as the following.

– Why is the plane represented by a parallelogram?

– Why are space coordinate systems represented with their first axis orientated downwards and to the left, their second one horizontally to the right, and their third upwards?

The inadequacies of any drawing can be brought to the fore (for instance, the drawing of a point leaves this point undetermined on a line), as well as the fact that, to be totally explicit, a drawing must necessarily be accompanied by a caption.

Besides, as has already been said, the demonstrations have required the use of a certain number of rules of space geometry (rules of incidence), and it is now time to institutionalise them, accompanying them of course, by drawings: some are assumed (axioms), the others being inferred from these (theorems).

5. CONCLUSION

So, by starting with the study of a most commonplace phenomenon, we have been able at one and the same time to introduce the students to space geometry, as well as to its representation. For them the rules of drawing will no longer be a sort of convention more or less implicit, that they have to discover by themselves and use intuitively (with all the risks that this entails, for coding as well as for decoding), but have acquired the status of listed geometrical properties, with no room for vagueness or for stereotypes, the problem of representation coming down to the choice of a good projection, that is a projection optimal for solving a given problem. Finally, extensions of these activities can be considered, on the one hand with regard to the different types of technical drawing, on the other hand with regard to art teaching, thus integrating mathematics with other academic subjects and using it to reinforce them.

REFERENCES

Audibert, G. (1985). Une problématique en géométrie de l'espace. Ed. IREM de Montpellier.

Bautier, T., Boudarel, J., Colmez, F. and Parzysz, B. (1987). Représentation plane des figures de l'espace, in *Actes du Colloque CNRS GRECO Didactique et Acquisition des connaissances scientifiques.* Sèvres. Ed. CRNS.

Colmez, F. (1984). La représentation plane en perspective cavalière des objets de l'espace, un problème de géométrie. Essai d'ingénierie didactique en classe de Première S, in *Actes du Colloque inter IREM Géométrie.* Ed. IREM de Marseille.

Parzysz, B. (1988). 'Knowing' vs 'Seeing'. Problems of the plane representation of space geometry figures. In *Educational Studies in Mathematics*, Vol XIX, No 1.

CHAPTER 13

The Exploration of the Space of Informatics and the Realm of Open Mathematics

B. Vitale
Laboratoire de didactique et d'épistémologie des sciences, FPSE, Université de Genève, Switzerland
Institut de psychologie, Université de Fribourg, Switzerland

ABSTRACT

The introduction of informatics into the school curriculum could provide, but does not automatically provide, a number of new and useful dimensions along which teachers and pupils could become sensitive to the logico-mathematical reasoning underlying the human construction of reality and could explore the realm of *open mathematics*, so seldom (if ever) present in school practice.

Several examples will be discussed which can be developed in a class with the help of qualitative, graphical and quantitative analysis and, where possible, real time graphics displayed on a computer screen. They will be used to outline the new integrated dimensions of natural and physical problems and their logico-mathematical components.

Special emphasis will be given to the intrinsic interdisciplinary character of most of the problems that can usefully be tackled by informatics at school.

1. INTRODUCTION

It is hard to conceive the relationships between mathematics and the real world when the very formulation of the query makes it almost impossible to be answered. It seems to imply that *mathematics* and the *real world* belong to separate realms and that some sort of a bridge ought to be found to join them; but then the bridge is generally metaphysical, found to join them; but then the bridge is generally metaphysical, making use of terms like beauty, harmony, simplicity and the like. It seems to me that mathematics is, as much as all other human activities, an *integral part of the real world*. The query to start from is probably quite different: what are the logico–mathematical aspects of our construction of reality? Is there any human construction of reality that could be void of abstract and formalised components?

I shall present here some considerations on the use of a very simple programming language (LOGOwriter) to explore empirical situations, where the curiosity about an adequate representation of a problem leads painlessly to the discovery of some of its logico–mathematical components (see also Vitale 1989a).

2.   PROBLEMS, REPRESENTATIONS AND FORMALISED MODELS

As will become clear in the examples that follow, each problem will present itself, at the beginning, as the result of a query about either an observation or an experiment – be it a real one, actually performed in the laboratory, or through an experiment that generalises a real one.

In order to state a *problem*, no formal tools are necessary; we can do it matter–of–factly, referring to the available empirical evidence, if any. However there is always, more or less hidden in what is presented as factual information, an interpretive framework that involves abstract and formalised thinking, even if school practice does not generally favour making explicit this logico–mathematical background. Once the problem is stated, we also have to find a suitable *representation* for it, be it in natural language, by gestures, by graphics, by numerical computation, and so on. This implies the need to select a suitable *representational space* for it and for its solutions: that will, in most cases involve a computer and the screen (Euclidian, two–dimension, vertical) space. We have also to choose a suitable *model* for it. This model will generally *simplify the problem*, in the sense of helping to define what is essential for us, and what is contingent in the phenomena, observed or foreseen. However, it will not necessarily *simplify the task*, as models can be as unpredictable and erratic as the natural phenomena they are supposed to model. As a matter of fact, models (or, in general, the whole of logical–mathematical thinking) are as much part of reality as galaxies, mountains, stones and trees, and they share with those objects the inexhaustible complexity of being (Vitale 1988a).

The problems, representations and models that I shall henceforth present, make use of the limited informatics tools available to secondary school pupils. Several of these problems are *open problems*, either in the sense that there is no unique correct solution, or that no (closed, analytic) solution is known at present. Teachers often assume that they should teach problems and related solutions only when the field has been sufficiently explored by professional mathematicians, so as to avoid unpleasant surprises and baffling phenomena. It seems to me, on the contrary, that *open problems should be a significant part of the mathematical curriculum.* In fact, most of the problems offered by everyday life are open, allowing only qualitative and approximate solutions. Only the simplest programming skills have been used in the projects outlined below. An effort has consciously been made to avoid recursive procedures, except for those making use of such a trivial brand of recursion as linear tail self-reference (Vitale 1989). Only a very limited subset of LOGOwriter primitives has been introduced and the emphasis has always been put on transparency of the procedures instead of on formal elegance and computing-time optimisation.

3.   OLD PROBLEMS IN A NEW GEOMETRIC OR ARITHMETIC GARB

3.1 Trees
This is a project born of an interaction between *drawing* and *informatics* (with a sprinkle of *logic* and *geometry*). Pupils can be asked to spend some time drawing trees. Then, by discussing with them their drawings and those of their comrades, the idea of a possible underlying *algorithm* can be made explicit. Without knowing it, many pupils follow a very simple algorithm: a trunk, then a fork, then two or more branches forking again ... (To avoid the imposition, by the experimenter, of the traditional and totally opaque recursive procedure for TREE, I have developed a different approach based on the creation of lists – positions and angles – to be read again at every construction step and to be progressively voided).

The main result is the beginning of a discussion, in an interdisciplinary context, of the relationship between *local rules* and *global configurations*, and of the effect of changing initial conditions (the starting point for the trunk) and numerical parameters (the length of the trunk, the angle between two branches, the number of branches at each fork, the number of branching levels) on the overall product. This theme will dominate, explicitly or implicitly, all of the projects discussed in this paper. (Trees can lead, mathematically, very far; see, for instance, Mandelbrot 1983).

3.2 Experimental psychology
There is no part of the *biology* curriculum devoted to *perception* nor, more generally, to *experimental psychology*. It is a pity, as so much could be understood about ourselves by experimenting; some such project could find a place in the optional parts of the curriculum. Perceptual *illusions*, however, are very hard to observe in a school laboratory, as the need to be able to modify dimensions, intensities, frequencies and the like could tax the laboratory much beyond its capabilities. Without loss of generality or efficacy, the computer could provide a cheap *experimental psychology laboratory*. Interesting *visual*

illusions, for instance, can be realised on the screen very well by quite elementary programming and graphical representations.   I think that we don't even need to speak of a simulation here, as the screen is as suitable as any other means to support the visual display of the illusion (see Perception 1985 for a whole set of relative motion illusions; almost all of the software provided in the enclosed floppy disk can be very easily programmed in LOGOwriter).

Relative motion, *Poggendorff* and *Muller* illusions can be discussed, programmed and individually tested to show the effects of the variation of parameters: frequency of flickering and distance between points (and, possibly, colour, intensity, number and form of moving points), in the case of relative motion illusions;   thickness and orientation of the bar and angle of the thinner line relative to the bar, in the Poggendorff case; length, distance and angle of the arrows, in the Muller case.   All parameters can be modified in real time, while the programme is running, by a trick allowed by LOGOwriter.   Here the curiosity is about the illusion effect itself, not the programming nor the geometrical considerations that make the programming possible.   However, a lot of careful geometric considerations have to precede programming, so that the presence of a logico–mathematical underlying structure in any realisation of the project can be felt easily.

I can only quote here a few other projects of the same type, for instance one on *music* (a project born of the interaction of *musical education* and *informatics*, with more than a sprinkling of *arithmetic* and *geometry* and a potential relation to *drawing*) and one on *plane mirrors* (construction of the virtual image).

4.   OPEN MATHEMATICS

4.1 Balls in a box

This is mostly a mathematical project, involving *chance*, *arithmetic* and *geometry*, but it can be used, as it will be seen, as an intriguing metaphor in biology and social sciences.   Take a box containing a white ball and a black ball: initial configuration [w b].   Then extract one of the balls at random; if it happens to be a white one then put back inside the box two white balls, if a black one then put back two black ones.   At the end of the first cycle, you have either a [w w b] or a [w b b] configuration inside the box.   The ratio r = (number of w):(number of b) can assume only two values, either 2 or 0.5 (Blackwell–Kendall urn, see Milgram 1983).

What will happen if you iterate the procedure a very large number of times?   When there are many white and black balls in the box, the change of one in the number of either whites or blacks should not influence greatly the value of r; so we could foresee (but this is still intuitive reasoning, disapproved of in probability theory) that r tends to some fixed value between 0 and infinity (extremes *not* included).   What if we start the game again? Shall we get the same r? A different r? With which distribution in the open interval (0,infinity)? The problem is a tricky one, as the probability of extracting a white ball or a black ball *changes with the number of extractions* in a stochastic way. This makes the standard branching–process method of probability theory inapplicable. Shall we go to the school laboratory and take a box and a lot of white and black balls and start extracting balls for a while? We can do it, but after a few hundred extractions teacher and pupils would be exhausted. Why use *physical* balls, though? Why not use just *symbols* on a screen: w, b? Why not just *dots* in a diagram, representing how r changes with the number of extractions? Why not different dots in another diagram, representing the distribution in r after, say, 10,000 extractions?

The biology and social sciences sideline: the (asymptotic?) value of r is strongly sensitive to those fluctuations that take place in the very first extractions. For an organism, for an individual: the first days, the first months of life?

4.2 Cellular automata and fractals

Cellular automata are a recent and almost inexhaustible source of models and metaphors in *physics* and *biology* (see Demogeot *et al* 1985, for details on the model and a recent review of applications to physics and biological sciences). A cellular automaton simulates the growth of an initial set of "cells", according to a given (and largely arbitrary) set of "growth rules". A particular two–dimensional version, *Life*, has proved particularly successful in modelling growth problems. The field is exceptionally *open*, as very little formal analysis has been possible up to now and the cellular automata have provided a rich field for *experimental mathematics*. Two–dimensional versions (in particular, Life) can be programmed in LOGOwriter and their evolution followed on the screen, but the pace is much too slow. However, the one dimensional version works very well and reasonably quickly, so that it can provide teacher and pupils with a new laboratory: the *mathematical laboratory*. See figure 1 for an example of a cellular automaton grown from an initial single living cell.

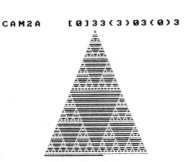

Figure 1

I can only cite here a few other projects of the same type: for instance, one on *population growth* and *population interaction* (see Vitale 1988) and one on stochastically constructed *fractals*.

5. CONCLUSIONS

Logico–mathematical reasoning is present, and necessary, in every human construction of reality.   In school practice of causal domains, it is sometimes hidden by empirical evidence, presented as conceptually neutral.   It is sometimes put aside by teachers as too difficult; it is very seldom made explicit, even when implicitly used.   When it is made explicit, it enriches our experience and gives us the possibility to pursue – at the level of extrapolation, generalisation and experimentation in new domains – the curiosity that phenomena and problems can elicit in the classroom.

Better for mathematics to be felt as an underlying presence in all of our thinking and as a unifying network in interdisciplinary projects, than as an esoteric field of pure reasoning.   This does not exclude the need for abstract and formalised spaces in the curriculum, nor the recognition of the specificity of mathematical truth.   It only makes any pedagogical approach to this truth an integral part of the construction of pupils' knowledge.

REFERENCES

Demogeot, J., Golès, E. and Tchuente, M. (eds) (1985).   *Dynamical Systems and Cellular Automata*.   New York: Academic Press.

Mandelbrot, B. (1983).   *The Fractal Geometry of Nature*.   New York: Freeman and Co.

Milgram, M. (1983).   Les formalismes du hasard.   In Dumouchel et al (eds) *L'auto–organisation de la physique au politique*.   Paris: Seuil, 182–209.

Perception (1985).   Human Motion Perception.   A special issue of *Perception*, **14**, 97–241; the issue contains an Apple floppy disk with the programs of the relative motion illusions.

Vitale, B. (1988).   Psycho–cognitive Aspects of Dynamical Model Building in LOGO; a simple population evolution and predator/prey model. *Journal of Educational Computing Research*, **4**, 229–253.

Vitale, B. (1988a).   The Unbearable Lightness of Models.   In Joffily, S. (ed) *The Physics of J Leite Lopes*.   Singapore: World Scientific Publishing.

Vitale, B. (1989).   Elusive recursion: a trip in recursive land.   *New Ideas in Psychology*, **6**, in press.

Vitale, B. (1989a).   Processes: a dynamical integration of informatics into mathematical education.   In Hoyles, C. and Noss, R (eds) *LOGO and Mathematics: Reasearch and Curriculum Issues*.   Cambridge (Ma): MIT Press, in press.

CHAPTER 14

Maths as a Human and Scientific Value in the Computer Age: a multidisciplinary approach centred on several problems of the Tagus River

M. C. Zambujo
Esc S Eca de Queiròs, Projecto Minerva, Portugal

ABSTRACT
The purpose of this paper is to present part of an ongoing study, taking place in a secondary school involving classes of 7th and 9th graders. The Tagus River has become a centre of interest in developing different kinds of activities, and some teachers of other subjects have also collaborated. Maths concepts have been developed, such as proportionality, equations, measures, matrices, geometry and statistics. The software used has been Open Access, MicroSoft Multiplan, Drawing, Graphing and Writing Assistant and an experimental Intelligent Tutor which contains knowledge about Arity/Prolog. The evaluation has been made through direct observation and achievement tests – one on maths concepts and the other on the interpretation of results – together with attitude reports completed both by teachers and students.

1.   INTRODUCTION
In recent decades we have seen a fast evolution in the new technology of information, making us foresee profound social, economic and cultural changes.

Computers are a reality, more and more present in our daily lives. The reformulation of an educational model considering the introduction of this new technology – the computer – is a long range process of intellectual, economic and social change. In my opinion, the question today is not whether computers should be used in schools, but to decide for what purposes and how to use them.

It is in this field that efforts are being made all over the world, and it is necessary for us, as teachers, to be aware of the important task we will have to perform. Obviously, new evolutionary and dynamic knowledge will be born while other knowledge will remain static and obsolete. The technology is connected with models of applications and problem solving. Thus, new models will bring about new technology and new technology will bring forth new models – with compasses we have one model of a circle, with the computer we have another (eg turtle geometry). The question is not 'Which is the best model?' but 'Which are the models that will be the most suitable under the circumstances?' and 'Which the closest to the reality of students?'. It is necessary to compare and relate these models.

2.   METHOD

The *aims* of the present study (a small part of a wider ongoing work – how to create pedagogical atmospheres in the computer age) are essentially to evaluate the potential of mathematics learning (in understanding real world problems by mathematisation, mathematical solution and interpretation, comparison of models using different technology; differences between seventh graders and ninth graders) and to explore the difficulties in bringing about an interdisciplinary environment where the computer is used as a tool for creating models and interpreting data, promoting investigation and the capability of phenomena interpretation and social skills.

The Tagus river (a nearby local river) has become a centre of interest in developing different kinds of activities because it is a source for studying aspects of local and community relevance, such as physiography, tides, variations of temperature, salinity and depth, samples of animals and plants. In short, it is a focus on cooperative work on real problems.

Themes were chosen such as pollution, tidal mills, salinity, temperature, species, substratum, depth, and so on. *Maths concepts* have been developed (such as functions, graphs, indices, percentages, proportionality, equations, measures, matrices, geometry and statistics), taking into account the process of mathematisation (discovering relations and structures, refining and adjusting models, using different models, generalising, etc). An example is given in section 4.

This study is taking place in Eca de Queirós Secondary School, Lisboa. It involves a class of seventh graders and another of ninth graders. These students are considered to be underachievers. The available equipment consists of PC compatible microcomputers, and the *software* used has been Open Access, MicroSoft Multiplan, Drawing Assistant, Graphing Assistant, Writing Assistant and an Intelligent Tutor which contains knowledge about Arity/Prolog made by a computer scientist collaborating in our study (Viccari et al 1987). All these materials are experimental.

Pupils are working in groups of three or four, and some teachers of other subjects have contributed to developing the activities. For instance, students have studied: in *native language* – polluting loads (carrying out interviews with the nearby population, making an excursion to the Olivais Dock and a sketch about polluting loads and analysing data by computer); in *geography* – cartography and remote sensing (using a colour code according to varying levels of pollution); in *French language* – the importance of Tagus history (making a report of Seixal excursion); in *history* – the importance of Tagus in the history of Lisboa, the archeology of Tagus valley, the importance of streams to the formation of cities and to the agrarian civilization; in *biology* – concepts of polluting loads, Tagus population, soils and substratum.

In this initial phase of the present study, we are more interested in discovering the general potential of the environment rather than isolating a problem and creating models. I think enough time should be provided to learn during and from experiences and studies.

The evaluation has been made through direct observation and achievement tests (formative tests) – one on maths concepts and the other on the interpretation of results, together with attitude reports completed both by teachers and students.

Pupils have also tried different methods of problem solving taking into account different technologies – they built graphs with paper and pencil and with the computer, they solved problems using different software and maths language, and so on.

3. AN INTELLIGENT TUTOR

We think that maths teaching should take into account all the possibilities offered by an investigation of Artificial Intelligence, which provide both development of scientific knowledge and investigation in the field of teaching/learning strategies. Among the languages of Artificial Intelligence available, Prolog (PROgramming in LOGic) seems to serve as the most generalised base of work as far as programming is concerned for the next generation of computers. Its interrogative power (query language) allows it to be used as an interpretative deductive language, making it possible for someone who uses it to maintain a conversation with the computer.

An Intelligent Tutor must build up the pupils' model, learn about the subject they are working on, and adapt to each student's style of communication. It must mix menus, natural language and graphic modes of interaction. To construct this tutor we focussed our attention on natural language communication, specifically on the tutor's ability to correct spelling, learn new words and their syntactic categories, new grammatical concepts and new syntactic rules. Learning and correcting depend on the tutor's hypothesis proposal and the student's agreement.

4.   AN EXAMPLE
Solving problems about the Tagus Estuary

Relationship between faunal species and soils
After copying to the spreadsheet one table containing types of soils in several sample stations of Montijo, and another containing the classification of soils taking into account the percentage of sand, pupils solved this problem.

If we collected ten faunal species in each station of Montijo to find a crab (Carcinus Maenas), how can we know the constancy of this species?

Students tried to define an index.   Thus the constancy index of species for each kind of soil was introduced as well as the fidelity index (the relationship between each constancy and the total of constancies). Also, they built this table in the spreadsheet containing the number of crabs gathered in each kind of soil:

Station	Number of Crabs	Type of soil
C01	1	VNC
C02	0	VNC
C03	7	O
C10	2	VNC
C11	4	AV
C12	2	VNC
C15	9	O

VNC – Black compact slimes
 O   – Oysterbed
AV   – Slimy sands

For calculating the constancy index, IC, in each type of soil, the usual math model was

$IC1 = 5/40 = 13\%$ 10 species in 4 VNC stations
$IC2 = 16/20 = 80\%$ 10 species in 2 O stations
$IC3 = 4/10 = 40\%$ 10 species in 1 AV station

Using the spreadsheet, in particular the pointer with VALUE function and FORMAT, for the percentage (%)

	Data			Formula
	4	5	6	6
37		VNC	13%	(RC(-4)+R(+3)C(-4)+R(+5)C(-4))/4
38		0	80%	(R(+1)C(-4)+R(+5)C(-4))/20
39		AV	40%	R(+2)C(-4)/10

Students then added these percentages

40	Tot		
41		133%	SUM(R37:39C6)

Of course the question appeared about the index constancy sum being different from 100% and the index fidelity sum being equal to 100%. The notion of percentage and proportion was discussed on this occasion.

Another group of students worked with PROLOG.

```
%index of constancy
ic(A,X):  info(A,N1,N2,N3),
X1 is N2*N3,
X2 is N1/X1,
X  is X2*100.
```

A - NVC, O, AV
N1 - number of crab found in each soil
N2 - total of samples gathered in each soil
N3 - number of stations with the same kind of soil

```
eg  %data
    info(vnc,5,10,4).
asking the computer:
we got:
```

?-ic(vn,X)
X = 12.5
yes

For the classification of the species, taking into account the constancy and fidelity index, the students using the spreadsheet worked with IF and the operator AND with VALUE function and copied the formula to the next two rows (38/39) with COPY DOWN. They asked one by one class (constant, common, etc) of species through an initial table of classification:

	Data	Formula
	11	11
37	#	IF(AND(RC(-5)>76%, RC(-5)<100%, "constant", "#")
38	constant	-----
39	#	-----

The PROLOG group used the following algorithm and found it more practicable and easier than a spreadsheet:

```
%Classification of species according to the constancy index

constancy(C,T,N,Y):-table(C,Y).

table(C,Y):- C=<12,Y=rare.
table(C,Y):- C>12, C=<25, Y=little_common.
table(C,Y):- C>25, C=<50, Y=common.
table(C,Y):- C>50, C=<75, very_common.
table(C,Y):- C>76, Y=constant.


C - constancy-index, T - Type of soil, N - species name,
Y - variable for the classification of species
```

In this activity students developed some mathematical concepts such as proportions, percentages, simple arithmetical operations, variables, equations, functions and vectors. They built models and tried them with different technologies.

5.   SOME CONCLUDING REMARKS

With regard to the use of the *Intelligent Tutor*, the interaction has helped the students to focus their attention, serving also as a motivation, in order to take an *explorative attitude* to the instructional material. The messages directed to each case (failure, alerts, triggers, automatic correction of errors) were received with surprise by students.   The Intelligent Tutor we have implemented so far is able to teach at different levels of expertise according to student's responses and to the evolution of a teaching session.

The *teachers of other subjects* engaged in this study wanted to continue it in the next year because they realised that students improved their knowledge and their social attitudes.   However, we often heard from some other maths teachers: "This content doesn't belong to the curriculum;   what a lot of numbers!" reacting initially against this study.

The students have shown understanding of the concepts approached and are enjoying participating in this study.   Ninth graders seem to apply the notions of interval and equations better than seventh graders. A lot of maths concepts were not known beforehand, such as commutativity and proportionality.   New concepts and new contexts appearing in this environment (for example, in the example of section 4, the constancy and fidelity index) gave meaning and power to mathematical ideas.   Also, what is perhaps the most important result, the students engaged in this study now care more about environmental, human and world problems in this technological society where people tend to think only about gains.

6.   REFERENCES

Viccari, R., Costa, E., Coelho. H. (1987).   *"A Prolog Tutor for Logic Programming"*, PEG, Proceedings of the Annual Conference, Exeter.

CHAPTER 15

Enhancing Probability Education with Computer Supported Data Analysis

R. Biehler
Institut für Didaktik der Mathematik (IDM), Universität Bielefeld, FR Germany

ABSRACT
Computers can be used in applied mathematics and science teaching to support modelling, simulation and the analysis of empirical data. These possibilities are analysed critically with regard to probability education at school level where data-free modelling is predominant. An initial impact of computers may be an increasing gap between probability and statistical data analysis, if case simulation is used in a limited sense. Some perspectives for using data analysis and theoretical modelling as complementary approaches will be developed. The use of models as reference frames plays an important role.

1. INTRODUCTION: MODELLING AND REAL DATA IN PROBABILITY AND STATISTICS

Combining the collection and analysis of real data with building theoretical models is a major opportunity of using computers in applied mathematics and science teaching (see eg Barclay 1988). It can be used to support students in actually *doing* science instead of a continued use of classical word problems, which also undermine an adequate understanding of the model concept. For probability education at lower and upper secondary school level, the relation between modelling and data analysis under the new technological conditions has to be explored further. Some thoughts on this will be developed in this paper.

A comprehensive approach to modelling in probability was seldom put into practice. This is partly due to complexity: a model for describing free fall is extremely simple as compared to a description for the complex structure of data in the most simple random experiment, for example coin tossing. In contrast to the free fall, people usually do not have experience with a long series of coin tossing. Moreover, comparison between model and data is often done informally in science (education). Comparison of probability models with real data is more complicated and the object of a full scale theory (statistics). This situation is reflected in the division of labour at the scientific level. Even if we look at a very model–conscious university course like that developed by Breiman (1969, 1973), we find an instructive separation. In the probability part, data–free modelling is practiced. Models are constructed relying on basic heuristics concerning the physical meaning of equally likely outcomes and of independent trials. This basic raw material is used to construct complex and compound probability models. In the statistics part, *initial* models are assumed to be given, and statistical methods are designed for this situation. A more or less *data–free* statistics is practised. A certain exception is testing goodness–of–fit, where the initial model is put into question. Breiman (1973, page 216) gives as a basic heuristic principle : "It is the departures from the hypothesized distributions that frequently give them most informative and interesting insights into the nature of the physical system producing the data". This attitude of using probability models as *reference frames* is also made explicit by Feller (1968).

On a limited scale, there is quite a tradition at school level in attempting to integrate ideas from statistics and model building into an (applied) probability course from the beginning, instead of reproducing the division of labour at the university level. All the activities offered to children with simple random devices such as games, spinners, dice and coin can fall into this comprehensive or holistic approach and comprise activities in all the areas. Working with random devices has pedagogical and practical significance because it provides opportunities for practical and cooperative work. It has epistemological significance because it supports developing the probability concept as a tool for handling random situations in real life.

2.   THE INITIAL IMPACT OF COMPUTERS

Although the computer might have been used for prolonging the latter approach, it had certain effects in the opposite direction. For probability education, the use of simulation is an obvious extension, which is already supported by a very limited technological equipment such as programmable hand–held calculators. Many practical suggestions for use in the classroom partly widen the gap between probability and statistical data analysis because:

(a) simulation is used to foster a sequential and dynamic interpretation of change experiments as *stochastic processes*, in contrast to the point of view of statistics, which prefers finite samples for inference and decision;

(b) simulation can be attractively used to extend the data—free modelling beyond the simple situations of current school mathematics, for instance including Markov processes or more complicated problems concerning simple situations (for example waiting times), which is mainly due to the assumption robustness of computational models;

(c) simulation is suggested for use as a partial replacement of real data, for instance by simulated Galton Boards and visualisations of the law of large number with artificial data (see also Biehler 1988b, 1989).

On the statistics side, there are complementary indications of further separation.   Statistics liberated itself, with the help of computers, from the universal control of probability and developed model and probability—free approaches, mainly for Exploratory Data Analysis (EDA). This has led to several novel approaches at the school level that aim at providing more independent room for data analysis instead of functionalising descriptive statistics for purposes of learning and teaching probability (see eg Biehler 1988a).   In practice, however, the methods of data analysis are also applied to improve model—based statistical analysis, ie to overcome the status that too unrealistic and grossly misleading models are used in statistics.   Data analysis can be used to develop initial models for a situation, to check model assumptions, and to analyse the structure of deviations (residuals) from a model used as *reference frame*.

Although these relations exist, data analysis and modelling can be regarded as different and partly complementary strategies for understanding a situation: a model may have a certain predictive power and may provide a simplified explanation and basic insight without being able to explain everything in the data.   Data analysis may reveal many peculiarities of interest without being able to provide such general insights as a model might do.   There are situations where only data analysis is possible, in other situations no data are available and modelling is a last resort.   This distinction between the level of describing relationships and the level of explaining them by theoretical models is common in science, but not in probability.

Simulation can play important roles.

(a) The extended modelling repertoire can be exploited to develop theoretical models for complex real data and study them by simulation.   The comparison between model and real data could then be done by comparing simulated and real data;

(b)   Data and model can be compared with simple graphical methods as
      developed by EDA.   If a more theoretical approach seems to be
      necessary in order to be able to judge whether a deviation could
      have been produced by chance, there is new room for developing
      ad hoc criteria and studying their characteristics under the relevant
      probability assumptions by simulation.   This approach is related to
      BOOTSTRAP methods.

This   new   flexibility   could   support   a   similar   flexible   and
complementary approach toward modelling and data analysis at the school
level.   A more or less normative modelling cycle, which gives each
element its place, may become very misleading.

3.   PERSPECTIVES AND EXAMPLES FOR USING REAL DATA IN PROBABILITY EDUCATION

The role of real data in individual applications of modelling and their
role in long term development of probability in the school curriculum are
two distinct problems.   For instance, it could be decided to offer a
course in Exploring Data and to rely on the manifold experiences made
available in such a course when probability is introduced later without
practising real data anlaysis so much at that stage.   In the following,
some basic options will be discussed.

3.1 Exploring and describing randomness

The new technological possibilities for exploring data could help to
introduce the probability concept as a *theoretical* concept (for this notion,
see Steinbring 1989) which is the basis of a theory which aims at
describing and understanding empirical phenomena.   We may illustrate
this with R. von Mises, who intended to rebuild probability theory as a
kind of empirical theory similar to mechanics.   The objects of probability
theory were conceived as 'collectives' that can be characterised through
two   fundamental   empirical   phenomena   (Urphänomene) :   stablising
frequencies and the principle of the excluded gambling system.   The
latter is closely related to the concept of stochastic independence and
expresses the complex structure in sequences of real random numbers.
There is a direct line from von Mises' foundational efforts to the
modern theory of random numbers.   On the other hand, school
probability has always *suppressed the complexity* in sequences of data
favouring the stablisation of frequencies.   Also, students will usually not
have experienced the many different manifestations of laws of large
numbers in reality, rather they will at best have seen one graph with a
make–believe sequence of real data.   Now (pseudo–)random numbers and
their structure have entered the neighbourhood of the school curriculum
together with simulation.   However, if the laws of large number shall
be established as a thought in reality, the first thing would be to use
computers to explore data bases containing real data from different

subject matter domains (insurances, vital statistics, etc) in order to support a rich knowledge of the generalised laws of large number. More generally, or philosophically, laws of large number can be interpreted as the existence of statistical regularities for a large number of cases despite irregularities and unpredictability in individual cases, as 'order emerging from chaos'. Experience with exploring data, where summarising, smoothing or middling data help reveal structure, can be related to this topic.

3.2 Modelling and reproducing variation in data

In the context of games of chance, the perspective that probability models are constructed to reproduce variation in data is not a natural one. Historically, it was the problem of measurement error that was a new challenge to develop probability models that were able to reproduce observed variation. This was more or less the beginning of a probabilistic research programme which aimed at conquering all kinds of processes with variation and uncertainty by using probabilistic methods.

In schools, using the Galton Board as a starting point for the development of probability lies on these lines. With computer support an extension would be possible. For instance, having more experience with real data on measurements, comparing the quality of different instruments for measurement (different variation), experiencing the existence of outliers and so on. Complementary, simple error generating probability models which support the idea that the observed error is the sum of many small independent errors could be actually studied by simulation, and the simulated data could be compared with data from real measurements. This would provide quite a different context than merely illustrating the mathematical central limit theorem by simulation. If students have had experience with data analysis concerning structures in bivariate data, in time series data, in sequential data (eg letters in texts) and the like, these may be taken up and be partly reproduced by simulation.

3.3 Modelling and data analysis as complementary strategies

If a library of data sets is available, word problems in probability could be enhanced with a more realistic context. A typical word problem in probability is the birthday problem: if s people meet in a room, what is the probability $p(s)$ that at least two of them have their birthday on the same day of the year? $p(s)$ can be computed on the basis of the assumption that each birthday is equally likely. The well known but surprising result that $p(s) > 0.50$ is true for surprisingly small s can be explained by combinatorical thinking. In a computer context, the birthday problem could be used as an opportunity for exploring how birthdays are really distributed over the days of a year. This could be extended to a small project in EDA. Smoothing or summarising data by month may help the analysis as well as plotting the residuals, ie the deviation between the relative frequencies and the expected 1/365. Results of several simulations of a uniform distribution could be graphically compared to the pattern in the real data, and p(s) could be calculated on the basis of a simulation taking the real distribution as the theoretical one. In the end one may have learned something new about patterns of human births as well as about the value of simplified models for explaining qualitative results, ie the high probability for a coincidence.

3.4 Comparison of model and real data

There can be many arguments for making a theoretical model experiential for students through simulation (see Biehler 1988a, 1988b). For instance, an interesting environment is provided by the program COINTOSSER INVENTION ANALYSER for the collection PROBABILITY AND STATISTICS PROGRAMMES. The basic idea is to use nine coin tossing machines that are implemented on a computer but where the models are hidden from the user. Each machine is either a fair coin tosser or represents a certain deviation from the ideal model of equal chance and independent trials (different kinds of dependencies, changing probability for HEAD, deterministic patterns, unequal chances for HEAD and TAIL, but independence). Students have several options for analysis and display of data for exploring the different machines and learning about what structure can be expected in random sequences. Such knowledge on theoretical models can then be used as a theoretical perspective for exploring real data. There are many interesting applications where the use of the model as a reference frame gives new insight into subject matter problems, for instance in process and quality control the expected structure of random series is an important reference frame.

If data libraries were available in probability education, this would be a rich source for showing students what kind of situations can be modelled by the standard distributions of the curriculum like binomial, Poisson and normal distribution. Nevertheless, the use of a reference frame may be the source of many more interesting applications, for instance the deviation from the binomial model concerning the distribution of male children in families with 5, 6, 7, ... children, how the balls in Galton Boards are really distributed and so on. As has been pointed out at the beginning, graphical methods may be of considerable help here, as well as simulations, which can be used for getting an idea on how large tolerable chance variations may be.

4. SOFTWARE REQUIREMENTS

Although there exist several software systems that support the modelling of dynamic systems and that would be usable in secondary education, such systems are lacking for the specific purposes of probability modelling. At present, array-oriented interactive languages like APL or

PC−ISP would probably have enough flexibility and options for graphical display, but they present problems with regard to user friendliness and ease of use.   It has to be further explored whether their extensibility and adaptability can be used to construct simpler learning and application environments for the purposes described above.   Statistical packages have to be carefully examined to see whether they have spreadsheet functions and flexible options for generating random numbers that may also support modelling, simulation and data analysis.   Such research is currently done in the SOMA (softwaretools for mathematics education) project at our institute.

REFERENCES

Barclay, T. (1988).   MBL to model: Combining real world data with theoretical models.   In Blum et al 1988.

Biehler, R. (1988a).   Educational Perspectives on Exploratory Data Analysis.   In Morris, R. (ed) *Studies in Mathematics Education Vol 7: Teaching of Statistics*.   Paris, UNESCO.

Biehler, R. (1988b).   Computer Simulation as Tool and Object of Teaching and Learning Probability and Statistics.   In Blum et al 1988.

Biehler, R. (1989).   Computers in Probability Education.   To be published as a chapter in Kapadia 1989.

Blum, W., Berry, J., Biehler, R., Huntley, I., Kaiser−Messmer, G. and Profke, L. (eds) (1988).   *Applications and Modelling in Learning and Teaching Mathematics*.   Chichester: Ellis Horwood.

Breiman, L. (1969).   Probability and Stochastic Processes: With a View Towards Applications.   Boston: Houghton Mifflin.

Breiman, L. (1973).   Statistics: With a View Toward Applications. Boston: Houghton Mifflin.

Feller, W. (1968).   *An Introduction to Probability Theory and Its Applications*.   New York: J Wiley and Sons.

Kapadia, R. (ed) (1989).   Chance Encounters.   *Probability in Education: A Review of Research and Pedagogical Perspectives on Probability in Education*. Dordrecht: Reidel, in preparation.

PC–ISP: Interactive Scientific Processor.   Artemis Systems.   Inc New York and London: Chapman and Hall; for IBM–PC and comp.

Probability and Statistics Program (1986).   Green, D. et al: Capital Media c/o ILECC, John Ruskin Street, London, SE5 0P2, BBC–micro.

Steinbring, H. (1989).   The theoretical nature of probability and how to cope with it in the classroom.   To be published as a chapter in Kapadia 1989.

CHAPTER 16

Using Real Applications in Teaching Probability Theory

P. Bungartz
Mathematisches Institut de Universität Bonn, FR Germany.

ABSTRACT
In order to get an insight in the usefulness of formulae and mathematical facts which pupils learn in so many lessons, and to get an insight into the necessity of mathematical proofs and theorems, we used real applications in our stochastic lessons. Motivated by the public discussion of nuclear energy and the risks of nuclear reactors, we tried to get qualified to criticise the official statements and argumentations. We had to learn the mathematical methods, the definitions and theorems which are used to compute the risk. Pupils got an idea of the part mathematical sciences are playing in the real world (Blum 1985).

The risk of nuclear reactors – a project for gifted students

1.   PRELIMINARY REMARKS

We started the project with a visit to the Research Institute for Nuclear Security (Kernforschungsanlage Jülich/Aachen) to get information about security problems. The students heard a lot of probability arguments and technical explanations. They didn't understand all this and had many questions. In our lessons we tried to reach understanding by studying the Deutsche Risikostudie Kernkraftwerke (TÜV, Köln 1980), the German interpretation of the Rasmussen, NC Reactor Study (USA, WASH-1400, NUREC-75/014, 1975). We only studied some chapters in order to understand the risk calculation and we looked at only one accident, the breakdown of the cooling system in the case of a leak in the cooling pipe of the core.   If there is no immediate cooling the reactor core will melt and the security casing can explode.   We didn't try to understand all the possibilities of a reactor accident and we did not intend to become experts in risk theory.

2.   TECHNICAL BACKGROUND

One student, a member of the physics course, explained the physical functions of a hot-water reactor, the nuclear reactions, the core, the security casing, the exchange of heat by two cooling circulations and so on.   In a special lesson we learned about the function of an emergency cooling system.   We had to explain the following technical diagram.

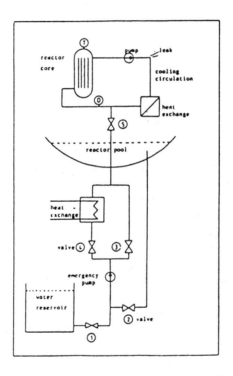

If there is a leak in the cooling tube of the core, valve (1) should open and pump P must start to pump cool water out of the water reservoir, past the opened valves (3) and (5), into the core-cooling pipe. The leakage is detected by measuring instruments D (pressure) or T (temperature).   This first cooling, flooding action, operates until the reservoir is empty.   Then the circulation action must start: hot water collected in the reactor pool must be pumped through the system. Valve (1) must shut the pipe, (2) should open, (3) has to close, (4) must open, cooled water (by heat-exchange) can pass the open valve (5)

into the core. The circulation activity must operate for several months.

3. FIRST CALCULATIONS

component	flooding action	circulation action	loss probability	
valve(1)	no opening		$p(v_1) =$	
valve(1)		no closing	$p(\overline{v}_1) =$	
valve(2)		no opening	$p(v_2) =$	
valve(3)		no closing	$p(\overline{v}_3) =$	4.0×10^{-3}
valve(4)		no opening	$p(v_4) =$	
valve(5)	no opening		$p(v_5) =$	
e.-pump	no start		$p(\overline{ep}) =$	8.1×10^{-3}
e.-pump		breakdown	$p(ep) =$	
heat exchange		breakdown	$p(he) =$	3.0×10^{-3}
instrument D	no indication		$p(D) =$	1×10^{-5}
instrument T	no indication		$p(T) =$	

If you work out the origin of these data you will discover the exponential and logarithmic Gauss distribution.

Assuming that all instruments and components work independently we make a first calculation of the breakdown probabilities:

P(breakdown of the flooding action) $= 2.012 \times 10^{-2}/a$
P(breakdown of the circulation action) $= 2.71 \times 10^{-2}/a$.

We can compute how often this happens if we know the relative frequencies of a medium-sized leak. Here $h(L) = 8 \times 10^{-4}$ per annum, so the probability of leak (L) and flood (F) is $p(L \wedge F) = 0.000016$ per annum. This number can be interpreted as happening twice in 100 000 years, so we can be sure that no accident will happen in our lifetime. This is wrong, and so we discussed the problem in depth.

4. REDUNDANCY

At the Research Institute we learned something about the technical efforts to reduce the loss probability by using the redundancy principle: instead of only one valve that should open a pipe in case of emergency, take two valves in parallel pipes. There is a breakdown only if both valves fail. The resulting loss probability for this arrangement is the product of the single loss probabilities, that means a remarkable increase of security or a reduction of the breakdown probability. We also get a multiplication of probabilities using connection in series of valves, if there is a pipe that should be closed in case of emergency. In our lessons we investigated the corresponding multiplication theorem of probabilities.

5.   ESTIMATION OF THE RISK

The risk is the number you get by muliplying a measure of the damage by the uncertainty.   The risk of a nuclear reactor is computed in the Risikostudie as the product of the number of deaths in case of an accident and the probability of that accident.   According to that study the risk number R = 10.1 deaths/annum.   This number is wrongly compared with the risks of employment,   7,800 deaths/annum, and of traffic, 14,400 deaths/annum.   This comparison should mean that you have a great risk in your private life but only a very small one caused by nuclear energy!   To understand the computation of the risk we worked up a table for teaching.   Using the real data of the Risikostudie we made some simplifications, considering only an average leak in the main cooling tube of the core and reducing our consideration of essential security systems.   We arrived at the following possibilities:

A1: The rapid switch out of the reactor core breaks down.
A2: The measuring instruments fail.
A3: The flooding action breaks down.
A4: The circulation action breaks down.
A5: The long-term cooling system doesn't work.

If there is a leak in the main cooling tube it should be detected by the instruments (A2) rapidly, in a mechanical way, the nuclear reaction should be stopped (A1) and cool water must be pumped into the core (A3).   After a short time the circulation action must start (A4) and afterwards, over a long time, the cooling system must go on working. Otherwise:

V1: The core melts and the security casing explodes caused by hot steam.
V2: The core melts and the security casing bursts (big/small leak).
V3: Breakdown of the security casing extractor – radioactive gas enters the atmosphere.
V4: Breakdown of the pressure regulation of the security casing, a leakage is possible after some time.

Each event causes a lot of damage.   We looked only at human deaths caused by leukemia or carcinoma.   According to the data of the Risikostudie we get the following table of conditional probabilities.

<u>Leak in the main cooling tube:</u>

frequency: $h(leak) = 4 \times 10^{-3}/annum$

		Breakdown prob of security sys (10^{-5})					Number of estimated human deaths
		A1	A2	A3	A4	A5	
		0.5	3	90	50	25	
Probability of the burst of the security box on the condition that a security system breaks down $(10^{-2}/\%)$	V1	10	3	2	2	2	T1 = 104 000
	V2	42	12	4	3	4	T2 = 60 000
	V3	23	25	21	18	13	T3 = 3 700
	V4	25	60	73	77	81	T4 = 1 500

This tabulation became the basis of the following lessons.   First the pupils had to translate it into a tree diagram and then they were asked to solve some tasks.

- What is the probability that, in the case of a leak in the main cooling system the switch off of the core fails and a steam explosion happens?

   $P_{leak}(A1 \wedge V1) = P_1(A1) \cdot P_{A1}(V1) = \ldots = 5 \times 10^{-7}$

   frequency $= h(leak) \cdot P_1(A1 \wedge V1) = \ldots = 2 \times 10^{-9}/annum$

- What is the probability that, in case of a leak in the main tube, the measuring instruments fail and the extraction breaks down?

   $P_1(A3 \wedge V3) = \ldots = 2 \times 10^{-4}$

   frequency $= h(leak) \cdot P_1(A3 \wedge V3) = \ldots = 8 \times 10^{-7}/annum$

- How often in a reactor year will the breakdown of the pressure control happen?

   V4 = (A1 and V4) or (A2 and V4) ... or (A5 and V4)

   $P(V4) = P(A1) \cdot P_{A1}(V4) + \ldots + P(A5) \cdot P_{A5}(V4) = \ldots = 1.26 \times 10^{-3}$

   frequency $= h(V4) = H(leak) \cdot P_1(V4) = \ldots = 5 \times 10^{-6}/annum$

– An engineer will find that this frequency h(V4) is much higher than other frequencies. He will ask for the reason. What is the probability that the event V4 was caused by the breakdown of the flooding action?

Using Bayes' formula we find   P(A3/V4) = ... = 0.52.

In the same way we can compute other posterior probabilities: P(A1/V4) = 0.001,   P(A2/V4) = 0.014,   P(A4/V4) = 0.30, P(A5/V4) = 0.16.   The most frequent cause of V4 is A3, so we must keep a close watch on the flooding mechanism.

Finally, we computed the average number of deaths if the core melts and ... .   We found this number as the sum over all parts T_i multiplied by the frequencies h(Vi) : $T = T_1h(V1) + ... + T_4h(V4)$ = 0.05 deaths/annum.   That is the risk of one reactor;   in Germany we have 25 reactors so that we get a risk number $R_{25} = 1.25$ deaths/annum.

In the Risikostudie you will find that the risk number of this damage (R_{25} = 10.1 deaths/annum) is nearly equal to our result, but we are missing a factor of 10.   However, if you look at the data and our simplifications you will see that we can easily lose a factor of 10.

On the other hand, all the data of the Risikostudie are only approximately correct.   In fact there should be a computation of the propagation of errors, which is not given in the Risikostudie.

The discussion of the meaning of that risk number of 10.1 deaths per year was very interesting.   In the Risikostudie there is one chapter concerning the interpretation of very small probabilities.   For example, if there is a probability of an accident $P(A) = 10^{-4}$, it is compared with the neolithic period which began 10,000 years before.   The implication is that the small probability means there is only one accident in 10,000 years, and nothing will happen in our lifetime!

Finally, at the end of the Risikostudie we find (p 121) a table in which, if we sum up the last column of data, we find the average probability of a reactor accident as $E(A) = 0.8 \times 10^{-4}$/annum.   This number is often mentioned in the news or on television to underline the security of nuclear energy.   However, it should be remembered that we have about 500 reactors in the world and if we assume that all reactors have the same degree of security and are working independently, we get an average of 1/20 per annum.   That can be interpreted as one accident (smelting of the core and blow out of radioactive gas) in 20 years! – remember Chernobyl!   On the other hand, a reactor year is no human year!   'Per annum' means working year of a reactor and, worldwide, we have now had nearly 5000 reactor years!

We have had very good discussions in our lessons especially on the statements of so called experts who often gave a wrong interpretation of breakdown probabilities or risk number.   Sometimes it was a political discussion because, obviously, experts try to manipulate public opinion.

6.   FINAL REMARKS

At the end of the project we went back to the Research Institut to clear up open questions.

- Are those data of the Risikostudie correct?
- How can we estimate the loss rates of technical things (pumps, valves, and so on)?
- Can we really estimate the number of deaths in case of an accident?
- Do you really mean the number $R_{25} = 10.1$ deaths/annum is the exact risk?
- Why are the interpretation of probabilities mostly wrong in the Risikostudie?

Not all questions were answered, but pupils got a very critical view of the 'mathematically obtained data'. Mathematically we studied the binomial distribution (switches, motors, ...), the Poisson approximation (breakdown of electricity), the exponential distribution (average of loss rates) and got a little insight in using the logarithmic Gauss distribution.

REFERENCES

Blum, W. (1985).   Anwendungsorientierter Mathematikunterricht in der didaktischen Diskussion. *Math Sem Ber*, **32**, 2.

Blum, W. et al (ed) (1988).   *Applications and Modelling in Learning and Teaching Mathematics*.   Chichester.

Bungartz, P. (1988).   *Risiko Kernkraftwerke*.   Reprint, Math Inst. University Bonn.

Bungartz, P. (1988).   Das Risiko Kernkraftwerke. *Mathematiklehren*, **29**.

Bungartz, P. (1988).   Die Badewannenkurve.   *Der Mathematikunterricht*, **9**.

CHAPTER 17

Mathematical Modelling in Distance Education – A Challenge for Teacher and Learner

M. L. Fuller
Capricornia Institute, Rockhampton, Australia

ABSTRACT

The question 'Can mathematical modelling be taught?' has been debated at length. A sequel to this question could be 'Can mathematical modelling be taught as a distance education course?' Apart from the work at the Open University, few mathematics educators appear to have become involved in attempting to answer this question. Indeed there are experienced teachers of mathematical modelling who would simply answer No. They argue that group interaction cannot be undertaken by the student studying in isolation. The Capricornia Institute, as a major provider of distance learning in Australia, has offered an undergraduate mathematical modelling subject in external mode since 1985. This paper will outline efforts made to compensate for the lack of the usual in-class group activities by the utilisation of a variety of support systems to achieve efficient modelling experiences for the external student.

1. INTRODUCTION

John Berry and Tim O'Shea (1984) in a paper on the distance teaching of mathematical modelling stated 'But teaching modelling at a distance is fraught with difficulties'. It certainly is!

The mathematical modelling course which they describe is one which is taught at the Open University – a course which is enhanced by professionally prepared audio cassettes and television broadcasts, and backed by tutors in the field.

Berry and O'Shea listed several particular difficulties which have to be overcome in teaching mathematical modelling as an external course. These difficulties include

(a) students can spend a considerable amount of non-constructive time on project work when guidance is not readily available,

(b) students do not have the opportunity to participate in group discussion of problems.

We will mainly concern ourselves with (b) in discussing the internal teaching of mathematical modelling at undergraduate level.

2.   MATHEMATICAL   MODELLING   AT   THE   CAPRICORNIA INSTITUTE

The aim of this paper is to describe the philosophy of approach for a course in mathematical modelling which is offered as a second year core subject in the Applied Science (Mathematics and Computing) degree at the Capricornia Institute, Australia.

Mathematical Modelling 2 is offered in both the external and internal modes.   Input from one group of students can be relayed to the other group − a useful input/output exercise.

Capricornia's students do not have the benefit of television broadcasts related to their subjects.

Tutors in the field are not always available.   In the case of this subject, tutors are not available except on once−a−semester routine outreach visits by a staff member who may not teach the subject.

Students are often isolated, thus local discussion group activities cannot be arranged for students enrolled in the subject.   The nearest student enrolled in the same subject may be 500 km away!

In the internal mode the course is run along the lines of the many courses in mathematical modelling described in the literature, especially the proceedings of the International Conferences on the Teaching of Mathematical Modelling and Applications (see the various proceedings, edited by Berry et al and by Blum et al).

Externally, there are problems and difficulties;   problems associated with ensuring that all students, especially those working in isolation, are given every opportunity to engage in meaningful modelling activities comparable to those available to students who are able to avail themselves of group discussions.

In general the students who enrol in Mathematical Modelling 2 in the external mode are professionals working in a particular field (engineers, experimental officers, administrators, computer programmers and high school teachers).   They can reside in a variety of locations in an area equivalent in size to all of Europe.

Students are given

(a) study guides and a project guide,

(b) resource books containing copies of papers which describe the modelling process across a wide range of disciplines,

(c) copies of project and assignment work carried out by students in a previous year of the course.

They have access to, and are encouraged to use

(a) teleconference facilities, with the link up of 6 or 7 students to discuss a previously assigned modelling problem or one of the models in the resource books,

(b) audio–cassettes which can be discussed in return by telephone on an individual basis,

(c) their own computing facilities,

(d) accelerated library loan procedures.

If students can be organised into small local groups this is done, and is possible in several of the larger cities.   We shall concentrate though on the problems of the truly isolated student.

3. ENHANCEMENT OF THE MODELLING PROCESS BY INVOLVING THE LOCAL COMMUNITY

If the isolated student works in a field where other professional persons are employed we encourage discussion and involvement with peer group workers.   This situation often leads to an extremely useful blend of academic modelling and industrial realism, as described by Hamson (1988) in his recent case for mathematical modelling.

Early in the semester, external students are urged to explore their local workplaces to ascertain if there are other persons employed locally who would be interested in participating in discussion and exchanges of ideas relating to that particular student's work.

The responses to this quest by external modelling students has been rewarding.

Internal students have made extremely useful contacts which have enabled them to discuss their modelling exercises and assignments with others who, although not enrolled in the subject, are interested because the subject is often relevant to their own work interests.   The input from other persons, employed or engaged in a diversity of activities, certainly has countered the problem of individual students working in isolation.

Examples of the type of contacts made are as follows.

(a) A high school teacher who made contact with staff working in the local mining industry.   This contact resulted in a project relating to modelling the conveyor system for moving steam coal in an open–cut mine in Central Queensland.

(b) An electrical engineer who set up discussion sessions with another employee who was studying graph theory as part of a degree course.   This liaison resulted in a model of the spread of influenza based on a graph–theoretic approach.

(c) A married lady, with involvement in the local farming community, who became interested in the design of farrowing pens for a local pig farmer.   Her project included an analysis of the dynamics of pig movements in a confined space.

(d)   A local government office employee who was able to interest a group of colleagues in the nuclear arms build-up and the associated literature relating to modelling the arms race.   The resultant input from the colleagues contributed to an informative modelling project entitled Can the Arms Race be Contained?

(e)   A silver-mine technical officer who worked, in conjunction with colleagues in his own workplace and with technologists from the Mine Research Centre located nearby, to examine methods for modelling the mine water treatment system with a view to making its operation more efficient, with emphasis on the quality of the local environment.

This sample is presented to indicate the type of local involvement which is possible.   In general, external students who have made the effort to make contact with other people in their local community, report that the exercise also benefitted the local contributors as well as themselves.   The crucial criterion for good liaison was that the modelling process was seen to be an interesting and useful method for looking at real problems.

The truly isolated student of mathematical modelling *can* have the opportunity for group discussion if contact can be made with interested local workers and other members of the local community.

The problems encountered by the student studying in isolation are well documented by Kahl and Cropley (1986) in their paper on the comparison of face to face and distance learning.   However the contacts described above show that involving members of the local community does help to overcome some of these problems.

The student who is the only one in the area enrolled in the modelling subject does encounter extra problems but, if that student is fortunate enough to be part of a work team, there are some advantages. There is opportunity to gain what Ole Skovsmose (1988) describes as reflective knowledge in mathematical education.   To develop reflective knowledge the educational process has to be organised differently from that which is usually related to the transmission of facts and skills.   The student usually has the opportunity to relate the modelling process to his own environment.   We also feel that we have made some progress in assisting the isolated mathematical modeller to achieve what Mogens Niss (1987) describes as the third phase of the applications aspect of the mathematics curriculum.   This third phase is where consideration of the field of application and the associated problems come before the introduction of the relevant mathematics.   Our external mode students are encouraged to define problems and list the problems associated with formulating a mathematical model.   They are encouraged to relate the subject matter to organisational processes in their workplace.

There are great challenges for the lecturer.

In general lecturers must act as course co-ordinator, field tutor and student mentor.   They gain experience in using audio cassettes and organising teleconferences.   Discussions with individual students on the telephone in their own home after work or school is an essential component of the external mode teaching and learning process.

In the future, staff involved with teaching mathematical modelling as a distance–education course will also assist students to seek out appropriate contacts in the student's local community, which may assist in providing a suitable discussion group activity and so compensate, to some extent, for the lack of in–class discussion group work.

4.   SUMMARY

There is no doubt that teaching mathematical modelling at a distance is a challenge both to the student and to the lecturer.   There is a need to motivate the student to switch from the experience of traditional learning of mathematics as facts and skills to a philosophical approach, involving orderly organisation of the problem, before attempting to apply mathematics.   This takes much discussion by teleconference and frequent input of material and discussion with local peers.

We feel we have made some progress in this transition for isolated students by extensive use of the following.

1.   Resource material involving carefully selected modelling exercises which illustrate both success and failure, including work submitted by similarly situated students from a previous year.

2.   Constant use of individual telephone contact and organised teleconference meetings not only between the lecturer and student but between students with similar isolation backgrounds.

3.   Promoting the student's workplace and local community as a source of experience and discussions relevant to the promotion of successful mathematical modelling.

4.   Prompt feedback by audio–tape, telephone and mail.

5.   Their own, or workplace, computing facilities.

The organisation for the lecturer is time consuming and at times frustrating.   However, the reward comes when a student working in isolation demonstrates ability in the art, science and craft of mathematical modelling by presenting innovative project work.

In the Aims and Scope preface to Avi Bajpai's *International Journal of Mathematical Education in Science and Technology* the first paragraph is an apt description of the difficulties encountered in teaching mathematical modelling externally.

"Mathematics is pervading every study and technique in our modern world, bringing ever more sharply into focus the responsibilities laid upon those whose task it is to teach it.   Most prominent among these is the difficulty of presenting any interdisiplinary approach so that one professional group may benefit from the experience of others".

REFERENCES

Berry, J. and O'Shea, T. (1984).   Mathematical Modelling at a Distance, *Distance Education,* **5**, 163.

Hamson, M. J. (1988).   The Real Case for Mathematical Modelling. *Bulletin of the Institute for Mathematics and its Applications,* **24**, 22.

Kahl, T. N. and Cropley, A. J. (1986).   Face to Face versus Distance Learning: Physiological Consequences and Practical Implications.   *Distance Education,* **7**, 38.

Niss, M. (1987).   Applications and Modelling in the Mathematics Curriculum – State and Trends.   *International Journal of Mathematical Education in Science and Technology,* **18**, 487.

Skovsmose, O. (1988).   Mathematics as Part of Technology.   *Educational Studies in Mathematics,* **19**, 23.

Stone, S. and Tait, J. (1987).   A Distance–Learning Modelling Project in *Mathematical Modelling Courses* editied by Berry et al., Ellis Horwood.

CHAPTER 18

Computational Modelling: a Link between Mathematics and Other Subjects

J. Ogborn
Institute of Education, University of London, UK

ABSTRACT

It is argued that the possibilities of computational modelling on microcomputers, especially with the aid of suitable modelling system software, can transform the traditional relations between mathematics and other subjects. There will be less need for careful mathematical preparation before certain topics can be taught in some other subjects; teachers and students in other subjects can become more mathematically autonomous, and the relation between the mathematical difficulty of certain topics and their closeness to reality will change. There may also be value in revising the approach in mathematics to the teaching of differential and difference equations, with greater earlier emphasis on numerical methods.

1. COMPUTATIONAL MODELLING

There are many ways in which the computer can be used to model real situations, the main kinds being perhaps statistical models, qualitative or rule-based models (for example using PROLOG), Monte Carlo models, cell automaton models, and dynamic (usually difference equation) models. This paper is not concerned with the first two, but concentrates on the last three, especially the last.

Monte Carlo methods have found some use in mathematics teaching, an example being dropping a needle onto a grid of lines to determine π. Models of this type do, however, have a much wider application. Typical of such applications are models of population dynamics, in which individuals of several species have probabilities of dying and of giving birth, these probabilities depending on their food supply which may itself depend on their eating others. A simple example would be sharks and fish in an ocean. Other examples include models of economic activity, of queues and of traffic flow.

Cell automata models are deterministic rather than probabilistic, but can also be used for problems such as those mentioned above (see, for example, Toffoli and Margolus 1987, Marx 1984). The model is expressed as the rules governing whether each cell in a grid will survive or die, or whether a new cell will be born at an empty lattice site. It is common that the rule is local, with the fate of a cell depending only on the state of its neighbours. The game of Life, devised by Conway, is one well known instance. Another important problem, soluble with such models, is the percolation problem, an instance of which is the propagation of a fire through a forest which has a certain density of randomly placed trees (see Marx 1987). Other instances include the conductivity of material made of conducting grains in a non–conducting substrate.

Such models have the character that the model (the probabilities, or the rules) is usually extremely simple but that they can generate, in a large population of cells, patterns of behaviour which are complex and sometimes very surprising. Models may well be nonlinear. They can easily possess various kinds of critical behaviour, for example phase transitions.

Dynamic models include two main classes: those representing numerical solutions of systems of differential equations, and discrete models using difference equations. Examples of the first include the decay of charge on a capacitor and the convection of air on a hot day, while examples of the second include the population of insects which breed at discrete intervals, or models of voting allegiances. Since the first are normally modelled with difference equations approximating differential equations, the distinction is not so much in the modelling methods used but in the nature of the variables modelled, and in the care which may be needed over methods of numerical integration.

It is clear that the class of dynamic models embraces a very large part of what is usually thought of as mathematics applied to the physical, biological and to some extent the social sciences, particularly economics and geography. It includes all that sometimes goes under the name of General System Theory. This, therefore, is an extremely important area of modelling, if one is interested in the relation of mathematics to other subjects.

2.   MODELLING SYSTEMS AND SIMULATIONS

It is generally speaking not difficult or laborious to program dynamic models directly in any computer language. The actual code representing the differential equations may well be only a few lines. Thus in BASIC

$$dQ = Q*dt/(R*C)$$

$$Q = Q - dQ$$

$$t = t + dt$$

is enough to define one iteration step in solving the decay of charge Q on a capacitor of capacitance C, discharging through a resistance R. Many simulations, presenting attractive screen displays of (for example) the ecology of a pond, the functioning of a nuclear reactor, or the manufacture of ammonia, contain not much more code than this to define the underlying scientific model.   The laborious work of programming is not here, but in the production of attractive and effective screen displays.   Whilst this work is important for professionally produced simulations, the work involved has little or no scientific content;   it is not what we would want pupils to learn.   By contrast, the short and often simple code representing the model inside the simulation is exactly what we need pupils to understand.   For these reasons, modelling systems have a place in teaching, as they do in scientific research.   Professional modelling languages for research help the scientist, technologist or economist to define, run and see output from their own models, without the need to program any more than the model itself.   Similarly, modelling systems have been developed for school and college use.

One such system, developed by the present author, is the Dynamic Modelling System.   It provides a screen editor for the user to type in code such as that above, using any variable names, and for the user to assign initial values.   Output can be seen in tabular or graphic form. The screen is split, so that different parts of the work in hand (eg model equations and graphic output) can be viewed simultaneously. Facilities for help, error messages, disk management and tutorial displays complete the system.

A more recent system, also developed by the writer, overcomes the restriction of the Dynamic Modelling System to graphic output from only one pair of variables at a time.   This is the Cellular Modelling System, which is based on a spreadsheet concept.   Cells, as in a spreadsheet, calculate numerical values, taking input as needed from other cells. Unlike a spreadsheet, each cell contains and displays four things:
-    the name of the variable computed in the cell,
-    the line of code to calculate the value,
-    any remarks the user inserts,
-    the current value of the variable.

The array of cells can be set to calculate cell by cell, or once for the whole sheet (as usual in a spreadsheet), or iteratively, calculating the sheet over and over again.   This last mode is the most common in using the system for modelling.   Users are encouraged to write models containing loops, an activity prevented by most commercial spreadsheets. Cells refer to one another by their variable names, so that a cell might contain:

No_of_decays
No_of_nuclei*const*dt
Note: const is decay constant
1000

with No_of_decays being the variable calculated, as the product of three values No_of_nuclei, const and dt, each defined in other cells. The value 1000 is the present number of decays. For variables needing initial values, the initial value can be inserted in this value slot.

The Cellular Modelling System provides graphical output by converting a cell to a graphics cell. Thus the screen can show graphically any or all of the variables in the model. A good example of the use of this facility is to watch a model of the decay of a radioactive species into a daughter product, which itself decays into a third stable species. The decay of the first, the initial build-up and ultimate decay of the second, and the slow build up of the third can be watched all at the same time.

A good spreadsheet with graphics, such as EXCEL, can, of course, also be used for modelling of this kind. One disadvantage is that cells must usually be referred to by arbitrary coordinates, not by name. Another is that recursive (looped) models are usually not allowed, so that many iterations must be done by calculating new lines of the spreadsheet. The main advantage of a spreadsheet is that teachers or pupils may have used it for other purposes, and so not have any new learning to do about how to use it.

Another, and very imaginative, modelling system is STELLA, originally devised for use with general system modelling. It is object-oriented, with models being constructed not with equations but by assembling and connecting icons on the screen. One main kind of icon is a tank, representing a variable that will be incremented or decremented during iterations. Tanks are connected by pipes, controlled by valves, which regulate the incrementing or decrementing of variables. A valve can be controlled by the values of other variables in the system. Levels of variables are indicated graphically, so that STELLA provides automatic graphic simulation. Cartesian graphs of variables can also be plotted, as can tabular output. Functions can be defined algebraically or graphically.

3.   EDUCATIONAL   ADVANTAGES   OF   COMPUTATIONAL/ DYNAMIC MODELS

Models of the kind just discussed have several important educational properties.   Perhaps the most important is that they are both simple and expressive.   Thus for example the model for a harmonic oscillator (in the first order Euler approximation)

$$F = -k*x$$

$$a = F/m$$

$$v = v + a*dt$$

$$x = x + v*dt$$

$$t = t + dt$$

says rather clearly and vividly that the return force F is proportional to the displacement x, that the acceleration a is determined by the force and by the mass m, that the acceleration adds to the velocity v in a time interval dt, and that the displacement is altered through the velocity.   Such a structure of thought is not so evident in an expression like $x = x_0 \cos(\omega t)$, nor even in $d^2x/dt^2 = -(k/m)x$.

In dynamics models, it is usually the case that each step in the calculation (in the Cellular System, each cell) corresponds to some simple physical process or to some simple definitional outcome.   Since the steps are simple and the structure is explicit, one can also look at the structure. By trying some other dynamics problems it soon becomes clear that only the first line, specifying the force, need change.   Writing

$$F = -g*m$$

in place of the force on a spring makes the oscillator model into one of free fall.   This is one way of making the general point that mathematically all dynamics problems have the same fundamental structure, however unlike their analytic solutions may be.

A further advantage is that real world complexities can be introduced often with rather little penalty.   Thus writing the force for the oscillator as

$$F = -k*x - d*v$$

introduces viscous damping into the problem, with damping constant d. The computer finds the solution no more difficult, even though the analytic solution is now appreciably harder to obtain.   Make the dependence of force on velocity more complex still (for railway trains there is a constant term and terms proportional to v and to v^2) and analytic solutions are out of reach of most school and many college students, but the computational solution is not.   We may illustrate this feature by a notional graph, of the difficulty of a problem against its reality.

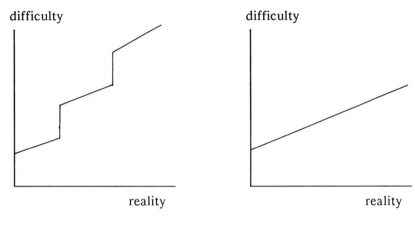

Analytic solutions   Computational solutions

The cliffs in the graph for analytic solutions correspond to needing more and more advanced functions to describe the solution. The smoother rise for computational problems reflects a somewhat greater complexity of the model equations, roughly in proportion to the amount of reality incorporated. Nor does the real world respect students' learning difficulties: if the oscillations of a drum require Bessel functions, then so much the worse for a student who does not know them.

A further educational advantage of computational models is that it is very easy to look at plots of various variables. Phase space is normally treated as territory best kept out of by many students, but plotting v against x for an oscillator is natural and simple. If the problem is, for example, one of the populations of a predator and prey, the seemingly irregular fluctuations of populations give way to an elegant and beautiful phase plot if the populations of predator and prey are plotted against each other. Here is another example of the computer changing the definition of what is difficult.

4.   CHANGES IN THE RELATION OF MATHEMATICS TO OTHER SUBJECTS

Changes in the relation of mathematics to other subjects which may result from wider use of computational modelling fall into two kinds: changes in how teachers and students of other subjects use and regard mathematics, and changes in the way teachers of mathematics need to relate to teachers of other subjects.

Different subjects differ quite markedly in how they define themselves in relation to mathematics. Physics teachers often tend, to the annoyance of mathematicians, to regard mathematics (at least calculus) as their own property. They think of Newton, Laplace, Lagrange and Hamilton as physicists, not as mathematicians. Many engineers think somewhat similarly. Such proprietorial feelings often make for difficulties with mathematicians, most notably when the latter provide service courses for physicists or engineers.

The use of computational modelling seems likely to reinforce rather than weaken such a situation, and to extend it to other subjects. Biologists, chemists, geographers and economists will have access to the possibility of 'doing their own mathematics' with computational models. The effects in such subjects may well be quite large. Inevitably, each subject develops a tradition of what is and is not possible in mathematics. The tradition is affected in part by the state of the subject itself, as recently in geography, with moves first towards and now away again from quantitative models. It is also affected by the kind of students the subject normally attracts, as for example in economics, where a long history of lack of mathematics has led to supply and demand equilibria being taught mainly graphically. So pervasive are such effects that they become part of the very definition of the subject. The dynamics of shifts of supply and demand linked through price, as a problem in coupled differential equations, is simply not part of secondary school economics. The existence of the possibility of computational modelling may therefore gradually change the definition of subjects.

Another kind of example can be drawn from the other kinds of models mentioned earlier: Monte Carlo and cell automata models. Here the problems addressed are, in the absence of such models, formidably complex, but are – given such models – able to be discussed with secondary pupils rather than exclusively with graduate students. One example is the use of Monte Carlo simulations in teaching statistical physics, where the Boltzmann distribution can be seen to arise directly out of the random exchange of energy, with no need for combinatorial algebra. Another is the simulation of competition between species already mentioned.

A different kind of consequence for mathematics in relation to other subjects arises out of the specificity of computational models. A model constructed, by an economics teacher, of flows of cash, profit and interest is likely to be seen by students as just that model, and not as an instance of something more general. Yet it may well be that models in chemistry, of flows of matter between chemical species, will have features in common with the first (for example the association of exponential change with linear first order systems). The danger is that mathematical insight may become fragmented; the corresponding potential advantage is that more students may more often think mathematically about more things.

This line of thought suggests a possible consequence for mathematics itself. In teaching differential equations, for example, the traditional order of events is:
– formation of derivatives,

–   analytic solutions of differential equations,
–   numerical approximations to solutions of differential equations.

One might suggest that the order could usefully be reversed, starting with numerical solutions to difference equations, without any formal calculus beforehand.   Then one might study the qualitative features of families of such solutions, so approaching analytic descriptions of the solutions.   After that, through studying the ills of finite difference methods applied to problems with continuous variables, one might finally define derivatives.   The progress in such a sequence is from the real and concerete towards the more abstract, from the mathematically particular towards the mathematically more general.   It need in no way be regarded as destructive of the mathematical passion for generality and form.   What it does is to work towards that instead of trying to start with it.   Starting with the abstract and general looks more mathematical, but runs the risk of presenting the real world as a clumsy disappointment, somehow inadequate to breathe the same mathematical air, instead of as a fascinating and endless source of problems which bring out all our creativity in thinking of ways to manage them.

In conclusion, let it not be forgotten that the real world is an important source of mathematics, and that this continues to be so in the computational era.   Newton made the calculus in order to compute the paths of the planets, not to give mathematicians interesting structures, though that is what indirectly he also did.   In our own time, the whole new and fascinating field of strange attractors in nonlinear systems, started with the meteorologist Lorenz trying to understand simple convection with the aid of a computer.   The reason the weather is hard to predict turns out to be mathematically very profound.   The particular often contains hidden surprises for the general!

APPENDIX

The Dynamic Modelling Systems, and the Cellular Modelling System, are both available from Longman's Ltd, Longman House, Burnt Mill, Harlow, Essex, CM20 2JB.   Both are available in versions for the IBM PC and compatibles, for the Nimbus, and for the BBC microcomputer.

The Cellular Modelling System can be made available with text on screen in various languages.

REFERENCES

Toffoli, T. and Margolus, N. (1987).   *Cellular Automata Machines; a new environment for modelling*. Cambridge Mass: MIT Press

Marx. G. (1984).   *Games Nature Plays*.   Roland Eotvos University, Budapest.

Marx, G. (1987). *Walking in the Plane.* Microcomputer software, Roland Eotvos University, Budapest.

CHAPTER 19

Practical Statistics in Other School Subjects

M. Rouncefield
Centre for Statistical Education, Sheffield, UK

ABSTRACT

Statistics is a useful and necessary tool required by large numbers of pupils for their project work in other subject areas. However, what are these pupils learning in their mathematics lessons to back up all this practical statistical work? What can the mathematics teacher provide to develop the necessary skills?

Often little consideration is given to the *quality* of the data collected and the most appropriate method of selecting a sample for a particular purpose.

Should the mathematics teacher consider sampling methods and help pupils to understand the need for a representative sample?

Can the microcomputer be utilised for a more practical approach geared to the child's understanding?

1. STATISTICS AND PROJECT WORK IN OTHER SUBJECTS

For the majority of pupils under the age of 16 in British secondary schools one major application of mathematics occurring in their other school subjects is statistics.

Pupils aged 14–16 taking a variety of optional subjects (including geography, sociology, applied science) now have to complete a project which involves collecting data, analysing data, presenting results in tables or diagrams and interpreting findings. Apart from a specialist examination in statistics, the techniques required by these other subjects are fairly simple graphs such as pie charts and bar graphs, and calculations such as the mean. Geography and biology also require the use of scatter diagrams for exploring possible relationships between pairs of values.   In addition, the humanities syllabus specifies that pupils must be able to construct questionnaires and present their findings by any means they think suitable (including charts and diagrams).

Statistics is clearly a useful and necessary tool required by large numbers of pupils for their project work in other subject areas. However, what are these pupils learning in their mathematics lessons to back up all this practical statistical work?   What can the mathematics teacher provide to develop the necessary skills?   First we consider the processes required in any project.

1.   The pupils must *specify the problem* and decide what it is they want to find out.
2.   They must decide *what data* is *needed* and whether and how it can be collected.
3.   *Collect the data* in an organised and efficient way.
4.   *Analyse the results.*
5.   *Interpret* the results and *communicate* the conclusions.

Unfortunately, the mathematics teacher tends, in many instances, to concentrate on stage four of the process and children may be drilled on quite complex statistical calculations without understanding their purpose. (The standard deviation is an example of this).   Often little consideration is given to the *quality* of the data collected and the most appropriate method of selecting a sample for a particular purpose.   I suspect that many children are going out with questionnaires containing badly phrased or leading questions which will give results which are inadequate, misleading or at worst entirely wrong.   Should the mathematics teacher consider questionnaire design and ways of avoiding bias at the data collection state?   How do pupils obtain their samples?   Is it satisfactory to take the first ten people they meet in the school corridor or at the local shopping centre?   Should the mathematics teacher consider sampling methods and help pupils to understand the need for a representative sample?

The points mentioned so far focus on the skills needed before any data collection is undertaken.   If we move on to the mathematics teacher's traditional domain, data analysis, we can consider the skills being taught in that area.   New techniques based on exploratory data analysis are beginning to be used in schools.   These allow the pupils to make informal inferences from the data without any complex calculations or statistical tests.

At age 18 the school pupil studying biology, geography or psychology is required to have extensive statistical skills in data collection and analysis including the use of hypothesis tests, rank correlation and X^2 tests (for both project work and examinations).   Pupils may find that they are being examined as much on their data interpretation and statistical skills as their specialist knowledge in geography, say.   Other subjects such as sociology and economics present pupils with numerical information which they must interpret.   On a specimen examination paper for the AS-Level Economics (Oxford Board) every question involved numerical data, four questions included quite large tables of figures, and one question presented information in a histogram.

2.   IMPLICATIONS FOR TEACHERS OF MATHEMATICS/STATISTICS

(a)   For all pupils below the age of 16 (and for older pupils studying sociology, history and economics) the main need is for simple skills in data interpretation, with careful consideration given to data collection methods.   All pupils should have a foundation course of this kind before undertaking any major piece of project work (of the statistical type outlined in this paper).

(b)   Older pupils studying geography, biology and psychology require these basic foundation skills plus more difficult techniques.   There is a need for more cooperation between teachers in these subjects and mathematics and statistical teachers.

(c)   Teachers in biology and geography may well feel overwhelmed by the amount and type of statistics they are expected to teach.   Many would welcome support from their colleagues in mathematics departments and from Local Authority Advisers.

3.   COMPUTERS – CAN THEY BE USED IN A GENUINELY PRACTICAL WAY?

At the data analysis level should the mathematics teacher concentrate more on simple graphical techniques, like those which are easy for children to understand and interpret?

Given that mathematics teachers may like to consider using more visual techniques and fewer complex calculations with pupils, what role can the computer play in this method of approach?

(a)   There are packages available which will draw stem plots and scatter plots. (Very few draw adequate box plots).   A useful example is Statistical Investigations in the Secondary School written for use by 11–16 year olds by Alan Graham.   This enables pupils to input data which can be instantly displayed on the computer screen in the format chosen by the child.   For older pupils Microtab will provide a variety of more advanced statistical techniques and significance tests, in addition to the simple techniques of stem plots, scatter plots and multiple plots.

(b)   A microcomputer can store data, so that the data set can be added to by different groups of children throughout the project, and diagrams or tables obtained to show the changing picture.

(c)   A microcomputer can be used to simulate results for large numbers of trials for a probability experiment, so that children can see the variation in the results of a random process.

In conclusion, I would like mathematics teachers to consider a less mechanical approach to statistics with more attention also being paid to the quality of data collected and the interpretation of results.

4. PRACTICAL STATISTICS – A CURRICULUM DEVELOPMENT PROJECT

In 1982 the Cockcroft Report included the comment that:

"Such textbooks (in statistics) as are available concentrate on theory rather than practice; it seems clear that there is a need for the provision of further teaching materials which will emphasise a practical approach to the teaching of statistics" (paragraph 780).

At the time, the Schools Council Project (directed by Peter Holmes) was already developing a series of practical statistics booklets for the 11–16 age group. This most recent project sponsored by the DES extends that approach into the sixth form.

Our brief for the project was to produce materials which could be used by students working to existing post-16 syllabuses, but we have also tried to anticipate new approaches to learning for this age group. So we have included sections on simulations and modelling, and also a chapter giving advice on how to devise and complete a longer investigation or project. We have tried to move away from a traditional approach of

Theory $\longrightarrow$ Examples $\longrightarrow$ Practice

to one of

Practical Activity $\longrightarrow$ Real data $\longrightarrow$ Discussion $\longrightarrow$ Model and theory

The emphasis is much more on active learning, with practical activity providing students with a basis of concrete understanding before they embark on the underlying theory. In addition the actual practical work itself will be much more memorable than any examples given in a text book or even found as part of a computer package (the internal workings of which may be totally mysterious to the student).

Our project materials include suggestions for data collection for all the probability distributions traditionally studied at sixth form level.

Each section begins by suggesting a problem such as the following.

* Can we devise a multiple choice test which candidates cannot pass by simply guessing the answers?

* Does the manageress of a petrol station need to install additional petrol pumps?

• Do girls and boys in the infant school make friends among both sexes or do they tend to segregate even at this stage?

• Can opinion polls predict accurately the outcome of the elections? If so, how are they able to do this?

The student is then directed towards the collection of data relevant to the problem. These are examples of the kinds of activities involved.

• A Frivolous Pursuit quiz requiring true/false answers is provided for students to try themselves and with a larger sample of pupils. By looking at the distribution of the number of correct answers the binomial model is built up from simple tree diagrams and the problem of the cheat-free multiple choice test is tackled.

• Pupils can observe and record the time intervals between cars arriving at a petrol station as a practical example of the exponential model. By considering the time taken to serve each customer, a simulation can be built up to explore the consequences of changing the number of petrol pumps.

• By observing the choice of friends sitting together at lunch time, a possible model for the random distribution of boys and girls at tables can be devised and tested.

• Pupils can administer their own opinion poll (possibly in conjunction with a mock election) and consider the problems involved in making a prediction based on a sample result.

Very little of the practical work outlined in the project materials is completely open-ended. Each practical aims to introduce a set of very specific skills and knowledge. However, an important by-product of practical work will be the increased confidence of students (and teachers) in their own ability to tackle a longer project or investigation. Indeed many of the practicals can be extended into longer projects or will spark off ideas for further exploration. The skills gained in short self-contained practical activites lay the foundations for project work, and there is a natural progression for the pupils towards a more independent way of working.

REFERENCES

Graham, A. (1987). *Statistical Investigations in the Secondary School.* Cambridge University Press.

Higginbotham, P. H. (1985). *Microtab.* Edward Arnold.

Rouncefield, M. and Holmes, P. (1989). *Practical Statistics.* Macmillan.

CHAPTER 20

Real Problem Solving in Mechanics: The Role of Practical Work in Teaching Mathematical Modelling

J. S. Williams
University of Manchester, UK

ABSTRACT
Three levels of real problem solving in mechanics have been identified, called intuitive–informal, ad hoc mathematical modelling and mathematical modelling within a scientific/Newtonian framework. Tasks involving practical activity in small groups using apparatus have been designed, which make modelling more accessible to students. They often form a bridge between two or all three of these levels. In particular, modelling with Newtonian mechanics is made more accessible to students by facilitating the formulation and validation stages.

1. INTRODUCTION AND MOTIVATION

The work described is connected with the Mechanics in Action Project. For background information see Collins (1988). Practical work was originally developed as a classroom starting point for Newtonian modelling, see Williams (1985, 1986/7 and 1988). Our work now encompasses the whole secondary range, and aims to develop further links with the science and technology curriculum.

The case for teaching mathematical modelling has been convincingly argued, in terms of motivating pure mathematics (Ormell 1975) and in terms of making the subject of applied mathematics coherent (Burghes and Huntley 1982, among others). In elementary mathematics, research by CSMS and by the APU has shown how the translation processes form a vital component in even simple arithmetic problems (see eg Hart 1981). Lesh (1981) has also argued the importance of modelling in strengthening average secondary students' mathematical skills.

The issue is therefore as Niss (1987) concluded in his survey on modelling in the curriculum, not 'whether' but 'how'. The question of the context of Newtonian mechanics is dealt with by Crighton (1985); he argues for teaching modelling in mechanics at sixth form and undergraduate level. There is certainly no shortage of potential curriculum material at this level in mechanics. Most recently, see the MEW group materials (in press). We therefore confine attention here to developing a teaching strategy for introducing secondary school students age 11 – 18 to modelling as real problem solving in mechanics.

We recognise three levels of real problem solving in mechanics appropriate to secondary schools.
1.   Intuitive–informal.
2.   Ad hoc mathematical modelling.
3.   Mathematical modelling within a scientific/Newtonian framework.

These are not seen as stages of cognitive development, but rather as modes of problem solving which become available as students mature and learn more mathematics and science.

2.   REAL PROBLEM SOLVING (1): INTUITIVE – INFORMAL

Many simple real problems are best tackled or solved without immediate recourse to formal mathematics or formal mathematical modelling. This can be a particularly effective starting point for younger (11–16) and average ability students, but experience shows that these starting points are useful for *all* students.

For example, given a collection of balls of various kinds, put them in order of bounciness, from least bouncy to most bouncy, and devise a procedure for measuring bounciness as a number from 0 – 100. Another example is, given a cane (with holes drilled), a nail, a bucket and a 100g mass, design and make or calibrate a yardscale.

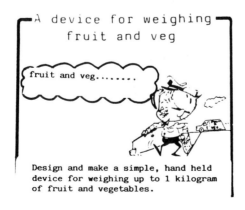

A device for weighing fruit and veg

fruit and veg........

Design and make a simple, hand held device for weighing up to 1 kilogram of fruit and vegetables.

Figure 1

Yet another example is, given string, springs and masses, make an instrument to measure time. (A stopwatch is provided to calibrate the instrument).

For many children, these activities involve considerable problem solving demands. The use of simple mathematics involving measuring, ordering, comparing and perhaps percentages or ratio, is situated in a challenging or worthwhile context. For many students, however, the problem posed can lead to extensive mathematical modelling.

3. REAL PROBLEM SOLVING (2): AD HOC MATHEMATICAL MODELLING

Many different flow diagrams have been proposed to describe the modelling process. For our purpose, the three stage diagram is simple enough but still retains the essentials (see figure 2). In ad hoc or empirical modelling we introduce variables, collect and represent data (stage 1), we then devise a mathematical model such as a graph, rule, pattern or formula (stage 2). This model has *predictive* power, and so the last stage is to make predictions and test them.

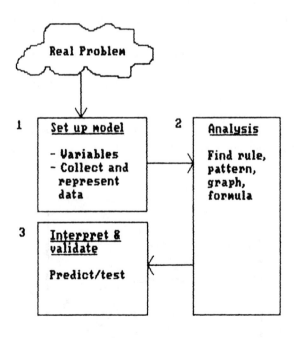

Figure 2

A group of 14 year olds, for example, investigated the relationship between the height of a ball before and after it's bounce (as a means of confirming whether or not their bounciness coefficient depended on drop height). Then they produced a graph and a formula and used this to predict rebound heights. Another group obtained a graph and formula for bounce time, $T = 6.8 \sqrt{h}$ seconds, where h was the height the ball was dropped from in metres.

Another class of 13 year olds drew a graph of the distance of the yardscale's markings from the fulcrum.

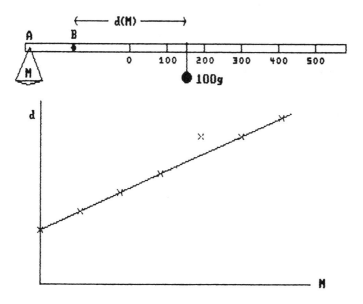

Figure 3

They interpreted the intercept of the graph as being the distance of the zero mark from the fulcrum. They concluded that one of their marks was 'off', since it was not on the straight line: this led them to go back to their machine and check the point again. Finally, they commented on their report that although d and m were not proportional, they could interpolate and mark off 10 grams per cm on the scale. (The distance between the holes A and B was 10 cm).

Several groups found they had to move the bucket nearer to the fulcrum to get the scale up to 1 kg, and this gave a steeper graph, but no one concluded that the distance between 0 and 100 would always be the length AB!

Ad hoc mathematical modelling of this kind is prevalent throughout applied mathematics and science. It is usually assumed that mechanics involves only classical Newtonian modelling. However, in fact, mechanics provides the opportunity for a great deal of empirical work in motivating pure mathematics and in introducing A-level mechanics.   Data collected on the amplitude decay of a mass spring oscillator involves curve fitting using an exponential function (figure 4).   These data were collected using a video, stopping the fast forward button every 2 minutes and measuring the amplitude.   Similarly, figure 5 shows how a cosine function x = a cos(ωt) + a was fitted to data collected from the video on slow forward, frame by frame.

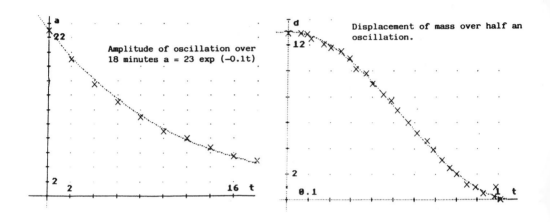

Figure 4   Figure 5

It is possible for this type of modelling to serve as a means of introducing new theory.   It provides an exploratory approach to such topics as elasticity, friction, moments, centres of mass, simple harmonic motion, impact and circular motion.

4.   REAL PROBLEM SOLVING (3): MATHEMATICAL MODELLING WITHIN A SCIENTIFIC/NEWTONIAN FRAMEWORK

Modelling within a well established scientific framework brings a qualitative change in range and power.   In the case of mechanics, the scientific framework is Newton's laws.   As Newton himself showed with gravity, one 'only' has to establish a model for the forces involved in a given situation and the application of Newton's laws of motion provide quantitative explanations of a whole range of phenomena, from orbits to the dependence of g on latitude.

In the case of modelling the motion of an oscillator, for example, model the mass as a particle of mass m, weight mg, the spring's tension as proportional to its extension (stiffness k), and ignore resistance forces. Newton's second law then gives

$$m \frac{d^2x}{dt^2} + kx = 0 \qquad (1)$$

where x is the displacement of the mass from its equilibrium position. The mathematical analysis (stage 2) now gives

$$x = a \cos(\omega t), \text{ where } \omega^2 = k/m \qquad (2)$$

where initial displacement determines the amplitude a.

We can now interpret this and validate it (stage 3) in practice. Measurements taken on the mass m with spring stiffness k always agree with this result very well, and the equation (2) clearly agrees with the video data in figure 5.

However, this model is only good for a limited time of, say 60 time periods (depending strongly on the type of spring used and degree of inaccuracy acceptable).

For longer time periods, a refined amplitude–decay model can be obtained from introducing an air resistance term into (1)

$$m \frac{d^2x}{dt^2} + \ell \frac{dx}{dt} + kx = 0 \qquad (1)$$

which gives

$$x = a \exp\left[\frac{-\ell t}{2m}\right] \cos(\omega t)$$

where

$$\omega^2 = \frac{k}{m} - \frac{\ell^2}{4m^2} \qquad (2)$$

Not only does this refinement provide a satisfactory model for the damping of the amplitude, which can be looked at practically by varying the mass and resistance force, it also predicts the new phenomena of critical and overdamping when $\omega^2 \leqslant 0$, which are of considerable interest to the design of car suspensions and galvanometers.

In general, the process diagram now takes on a modified form (figure 6). The assumptions in stage 1 usually now involve selecting some stock models for the bodies and forces involved. For instance, bodies are often modelled as particles, or rigid bodies. Strings are often assumed light. Resistive, tension or friction forces are commonly modelled as linearly dependent on an appropriate parameter, and so on. Practical work can give meaning to these stock mechanical models. Consider the problem of what happens to the 10 gram masses on the end of a (real) string over a (real) pulley, as shown in figure 7. What happens when the pulley is released? Is the pulley light and smooth? Is the string light? Suppose you wet it? Are the masses identical? Suppose the masses were taken off, what would happen to the string?

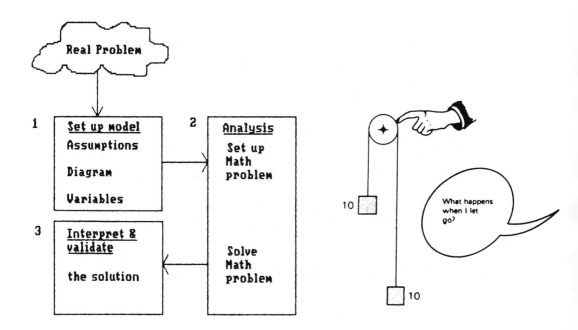

Figure 6   Figure 7

The nature of the stock models of forces can similarly be made meaningful and accessible through practical work. Hooke's law for a spring can be investigated for a variety of springs and strings (using ad hoc modelling again) with interesting results. Students are sometimes shocked and sometimes delighted to find how the model is very limited for most real springs and elastic strings.

Stage 1 also involves selecting appropriate geometry and variables for the problem. This alone involves sufficient flexibility usually to provide scope for variety of approach and individual creativity. (See Berry, Savage and Williams, in press).

5.   THE ROLE OF PRACTICAL WORK

To see the dimension which practical work can add to the whole modelling process in mechanics, we consider an example of a problem trialled with 15–17 year olds. The problem is to design a roadblock with a suitable counterweight (figure 8).

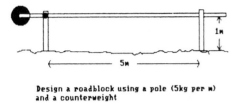

Design a roadblock using a pole (5kg per m) and a counterweight

Figure 8

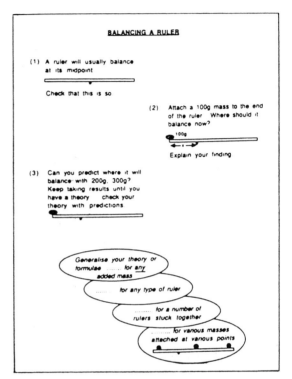

Figure 9

We use a ruler balancing practical (figure 9) to help students get to grips with the basic concepts. The ruler is a half-way model of the pole, the 100 g masses model the counterweight. The student can fall back on ad hoc modelling to obtain graphs (figure 10) or formulae, for the relationship between the mass and the balance point. (This is an exercise which has been successful also with 11 and 14 year olds). However, modelling with the principle of moments now provides a general formula

$$d = \frac{ML}{(M+m)}$$

where the length of the rule is 2L and its mass is M. This is a powerful result which helps in the design of the roadblock, although there are other issues involved, such as the force the guard should apply to lift the roadblock, and the contact forces at the points of support.

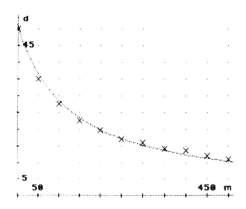

Figure 10

The strategy here is to provide practical work which relates closely to the real problem.   This usually provides a means of helping students to come to grips with the formulation of a suitable mathematical model for several reasons:

(a)   the apparatus provides a simplification of the real problem in itself − it isolates a few of the problem's essentials;

(b)   the practical allows the students to interact with the problem in the classroom − they manipulate the apparatus, discuss the principles and critical variables, and so on;

(c)   the apparatus allows the students to define a set of real problems which can be more easily solved and validated before tackling the associated real problem;

(d)   the apparatus can often provide data which can lead to ad hoc modelling of relationships without the prerequisite of Newton's laws, but which is consistent with, and can lead to, Newtonian modelling.

In this way our practical approach forms a bridge from level 2 to level 3 modelling and it makes Newtonian modelling accessible to more students.

6.   CONCLUSIONS

Our strategy in teaching modelling in the context of mechanics can therefore be described as designing tasks and practical activities which:

(a)   connect with real problems,

(b)   allow hands−on experimentation with apparatus in the classroom,

(c)   encourage real problem solving at all three levels described

(d)   make the stock models of classical mechanics meaningful to students as approximate models (often in need of refinement).

The   intervention   of   the   teacher   can   be   seen   as   an   attempt progressively to develop students' modelling capabilities to include the application of mathematical knowledge, and to include progressively the three levels of problem solving described above.

REFERENCES

Berry, J. S., Savage, M. D., and Williams, J. S. (in press). Case Studies in Modelling in Mechanics. In *Teaching Mathematics and its Applications*.

Burghes, D. N. and Huntley, I. (1982). Teaching Mathematical Modelling, Reflections and Advice. In *International Journal for Mathematical Education in Science and Technology*, **Vol 13**, No 6, 735, 754.

Collins, W. D. (1988). The Mechanics in Action Project. In *Bulletin of IMA*, **Vol 24**, No 1/2, January/February, 2–6.

Crighton, D. G. (1985). Why Mechanics. In Orton, A. (ed) *Studies in Mechanics Learning*. Leeds University Press.

Hart, K. (1981). *Childrens Understanding of Mathematics: 11–16*. London: John Murray.

Lesh, R. (1981). Applied Mathematical Problem Solving. In *Educational Studies in Mathematics,* **12(2)**.

MEW Group (in press). Exploring Mechanics. London: Hodder and Stoughton.

Niss, M. (1987). Applications and Modelling in the Mathematics Curriculum. In *International Journal for Mathematical Education in Science and Technology,* **Vol 18**, No 4, 487,506.

Ormell, C. (1975). Towards a Naturalistic Mathematics in the Sixth Form. In *Physics Education*, **10**.

Williams, J. S. (1985). Using Equipment in Teaching Mechanics. In Orton, A. (ed) *Studies in Mechanics Learning*. Leeds: University of Leeds.

Williams, J. S (1986/7). Practical Applied Mathematics, Part 1, Part 2. In *Mathematics Teaching*, (1986, p116) and (1987, p118).

Williams, J. S. (1988). Practical Mechanics: Mechanics in Action. In *Mathematics in Schools*, **17**, No 2.

CHAPTER 21

Experiences with the Numeracy through Problem Solving Project

J. Gillespie and B. Binns
Shell Centre for Mathematical Education, University of Nottingham, UK

ABSTRACT

Many students in school have little opportunity to use their mathematics in solving extended real-life problems and putting their solutions into practice. A team at the Shell Centre for Mathematical Education, University of Nottingham, has been developing classroom materials to support such activities under the title Numeracy through Problem Solving.

This paper summarises the basis of the project's approach. It then includes extracts from interviews and comments recorded by the team during the trials of the project materials. The two interviews are with groups of lower ability 15 year old students engaged in trialling the module Plan a Trip, and have brief commentary notes. The interviews serve to illustrate ways in which the project aims were realised.

1. INTRODUCTION

For the past four years, a project team based at the Shell Centre, University of Nottingham, and the Joint Matriculation Board, a schools examination board based in Manchester, has been developing classroom materials and associated assessment procedures under the title Numeracy through Problem Solving.

An important aim of the project has been to enable 14–16 year old students of all abilities to pose, solve and evaluate their solutions to extended real-life problems. This aim is in tune with key statements in the Cockcroft Report, itself at the centre of many of the developments in mathematics education in Britain at present, including

"Most important of all is the need to have sufficient confidence to make effective use of whatever mathematical skill and understanding is possessed, whether this be little or much." (paragraph 34)

and

"Our concern is that those who set out to make pupils 'numerate' should pay attention to the wider aspects of numeracy and not be content merely to develop the skills of computation." (paragraph 39)

This paper contains extracts from some of the interviews carried out by the team in the course of development of the project. Details of the associated assessment scheme are summarised elsewhere (Gillespie 1988).

2. CONTEXTS

The team sought problem contexts which would
* be attractive for students to work with,
* not make unreasonable demands on schools and teachers,
* have end-points or products which the student sees as worthwhile targets,
* develop strategic skills in areas including planning, organising, designing and choosing,
* encourage students to use simple technical skills effectively.

3. MODULES

To date, five modules of classroom materials and detailed teacher support have been produced, each corresponding to 10-20 hours of class time. A feature of each module is the importance attached to students working in groups, explaining their ideas and listening to each other, making their own decisions and learning from them, just as they do in life outside the classroom. While working through the modules, students themselves become responsible for setting and tackling their own problems rather than simply responding to tasks set by the teacher.

The modules in the series are as follows.

1. *Design a Board Game* in which groups of students design and produce a board game, which can be played and evaluated by other members of the class. The students spot faults in other games and seek to produce original and better games themselves using geometric, planning and other skills.
2. *Produce a Quiz Show* in which students devise, schedule, run and evaluate their own classroom quiz. Here simple statistics and probability along with timing and planning skills are important.

3. *Plan a Trip* in which students plan and undertake one or more class trips, and possibly some small group trips. Students are naturally involved in working with time, distance and money as well as in estimating, using timetables, obtaining travel information, and so on.

4. *Be a Paper Engineer* in which students design, make and evaluate 3-dimensional paper products, such as pop-up cards, envelopes and gift boxes. Here the ability to visualise and work in 3 dimensions is developed, to a degree of complexity dependent on the student's ability.

5. *Be a Shrewd Chooser* in which students research and produce consumer reports and advice on specific groups of products. Students experience product sampling and hypothesis testing at first hand.

Since students are responsible for selecting mathematical techniques which they can use most effectively and which they see as most appropriate, it is not possible to say for certain which will be used by every student.

However, through the five modules, all students use a wide range of techniques and concepts which include:

the ability to
- carry through simple calculations with suitable accuracy, using a calculator where appropriate,
- make estimates,
- make measurements (including number, length and time),
- draw accurately,
- interpret and display data in a variety of representations (including graphs, maps, timetables and other tables);

understanding and using some techniques of
- probability and statistics, including ideas of fairness, bias and randomness,
- ratio and proportion in many quantities, including ideas of enlargement and sharing,
- geometry in two and three dimensions, including working with parallelograms, simple rotation and construction of solids,
- logical reasoning, including the ability to enumerate alternative possibilities and classify them, and to select a best compromise solution to a complex problem,
- research skills, including the collection and evaluation of relevant data, using the telephone and using several sources of information in turn.

The planning process underlying each module has emerged in four stages, typically

1.   familiarisation,
2.   formulation of a problem and generating a possible rough plan for its solution,
3.   detailed planning and preparation,
4.   putting the plan into action and evaluating it.

Thus for Plan a Trip, we see the following stages.

Stage 1:   groups use a card game to simulate imaginary trips, encountering problems and seeking to put them right by better planning.

Stage 2:   groups discuss alternative destinations and means of travel for a class trip and make a final choice.

Stage 3:   the class lists and then allocates preparatory tasks to groups, who then carry out the tasks before the trip takes place.

Stage 4:   the trip takes place; afterwards the students reflect on what happened.

Since the students are involved in seeing complete problems through from beginning to end, they have to use a much wider range of skills than those often seen in the mathematics classroom, with a considerable overlap into other subject areas. This is in contrast to the more normal situation, where data is presented to a class by the teacher, and where problems are rarely tested out in practice. Social skills are particularly important within the student groups, in obtaining information from outside the classroom and in communication with teachers, other students and adults.

4.   OTHER CONTEXTS

Groups of students suggested many other interesting and worthwhile themes; each could form the basis for a further module. These include

planning and running a jumble sale, raising money for charity through sponsored events, planning and running a magazine, setting up a small business, planning a party, designing a bedroom, planning a youth group weekend, making a garden, orienteering, designing and making T-shirts.

Several of these have already been tried out by individual classes both in mathematics class time and in time set aside in school for more general activities, and have stimulated much worthwhile work.

5.   OBSERVATION

Owing to the novel nature of module activities in the mathematics classroom, the team has placed much importance on observing trials of the materials in the classroom.   These observations have formed the basis for modifications of a subsequent trial cycle; typically a module has passed through four or five such cycles before reaching the final published form.

The observations have included discussions and interviews with groups of students as well as with their teachers.   Extracts from these appear below, with brief explanatory notes on the right.

Extract 1.   Part of an interview with four lower ability 15 year old students in stage 3 of Plan a Trip.   Haywood School, Nottingham, July 1986.
(I is the interviewer, the others are the students in the interview group).

I   So where are you going?

S1   Skegness   Skegness is a seaside
                                                  resort about 60 miles
I   What have you done so far?   (100 km) from
                                                  Nottingham.
S1   Made a big list out, seeing what
      jobs we've got to do;   then we
      tick which ones we've done so we
      know if we've missed any out.
      We've got to phone the coaches
      up ... then there's the trains.

I   Have you talked it over with   Explaining or justifying
      your parents?   the activities to parents,
                                              developing social skills
S1   My mum says "It's interesting   and self-confidence.
      but why don't you get on with
      proper maths?"

S2   I told my parents.   They think
      it's good ..... it's more
      everyday maths and if I'm
      learning something they don't
      mind.

S2   It gives you more confidence.

I   Would you think of doing something
      like this ... planning a trip
      yourselves in the holidays?

S2   Yes. There's two people in our
     year planning to go youth hostelling.

S1   But if we went it wouldn't be   Transfer of planning
     planned like this. We'd just   skills – has this taken
     go!   place yet?

I   Is it worth planning, then?

S1   Yes, because if you don't,
     you're bound to forget something!

S3   It's more organised.

I   Have you talked about how   They have found out
     much it will cost you?   about a special group
     (the trip to Skegness)   ticket price for the
                                         train – cheaper but less
S2   Yes it's £2.35 by train ...   convenient than a hired
     there's two free tickets   coach for their class
     for staff or two others.   of about 25.

I   So what about the bus?

S3   £5.29 each by Camm's Coaches.

S1   There are ordinary buses to
     Skegness. We've got to phone
     up to find the cost.

I   How do you feel about doing   Exploring the students'
     this as a maths lesson?   attitudes to the activities.

S3   Some say it isn't proper maths
     but we get on with it. I like it!

S1   It's still connected with maths
     anyway. You've got to find out
     how many people have to go ...
     how much money you'll need for
     travel expenses ... it's more
     everyday maths.

S4   We've just been phoning up   Still the cost per person
      coach companies and the   is too high.
      cheapest was £100. We've found
      one for £90 for a 20-seater.
      Some are too expensive. Some
      are not there any more. Gone bust
      or something!

I   Would that be worth considering ...
      the £90 coach?

S4   We'll need a calculator!

S1   There's two maths groups ...   One of the group suggests
      some may come from the other   a way of reducing the
      class ...   cost per person – hiring
                                          a larger coach and
                                          including students from
                                          another class.

Extract 2.   Part of an interview with a group of low ability 15 year old
students after their class had organised and been on a trip to a Leisure
Centre and Country Park.   Greenwood Dale School, Nottingham, July
1986.

I   Was it all right to do it in a maths lesson?

S5   You can learn to plan your own   Students support the
      trips, can't you, learn more   novel activities in their
      things?   mathematics lesson.

S1   It was all right, but a bit weird.

S2   You still need to do calculating –
      how much money it was – how to get
      there, using timetables and
      everything.

S1   But next time we could go a bit   They see their trips as
      further couldn't we?   forerunners of more
                                           ambitious trips.

I   Would you plan another trip?

All   To London.

I   You reckon you could do that.

All   Yes.

I   Does anyone think you shouldn't
       do this in maths? People might
       say you should be doing proper
       calculations in your books.

S6   Yes but you've done that haven't you.
       Money, planning the times.

Extract 3.   An unsolicited letter from a teacher who had been involved
in trialling materials with her class.   This illustrates the change in her
student group (and in herself) as her students take responsibility for their
decisions and the outcomes of their planning.

> "I found it very difficult not to intervene and direct work to begin
> with.   It took me a while to realise what I intended as open-ended
> suggestions like 'have you thought of trying ...?' were simply
> interpreted by the students as 'Do this!'   It was difficult to stand
> back and allow them to make mistakes or explore blind alleys.
>
> The students initially complained that we were not doing
> proper maths but as they became more involved in the modules and
> discovered they enjoyed doing the work, the complaints became less
> frequent.   Now after four modules I think they are beginning to see
> that in fact they have developed some very valuable skills through
> the work and that it may be of more use to them than proper
> maths.
>
> Using the Numeracy Project has had a significant impact upon
> my relationship with the class.   I am sure this is because they have
> taken responsibility for their own learning and have had to make
> and live with decisions which have affected the whole class.   I have
> certainly discovered talents and personal qualities in many of them
> which would never be revealed in a traditional maths classroom.   A
> below average maths class has been transformed into a very
> impressive and interesting group of young people – or perhaps they
> always were?"

Through formal and informal conversations and observations, it
became clear that students were beginning to use skills in the classroom
which they had previously thought of as more to do with life outside
school.   Students who had been seen in school as of low ability now
displayed social and organisational skills of a far higher order than
expected – skills which they were used to using only in life beyond the
school gates.   In other words they were beginning to use their
mathematics as a genuine help in solving problems they saw as
worthwhile.

There have been good indications from trialling to support the view that classes of students who have undertaken several Numeracy through Problem Solving modules are prepared and keen to use their new-found skills and confidence to tackle more ambitious projects with good chance of success.   Many have spoken of their new-found and very useful skills.

REFERENCES

Cockcroft, W. H. and others (1982).   *Mathematics Counts.*   Her Majesty's Stationery Office.

Gillespie, J. (1988).   *Assessment of Mathematical Modelling.*   Theme Group 4 ICME 6.

Swan, M. B., Binns, B. E. and Gillespie, J. A. (1987–1989).   *Numeracy through Problem Solving.*   Modules and Teacher's Guides, Shell Centre for Mathematical Education and Longman Group UK Limited.

CHAPTER 22

Problem Solving for First Year University Students

F. Grandsard and A. Schatteman
DWIN, Vrije Universiteit Brussel, Belgium

ABSTRACT

During the past two years we have offered a problem–solving course to our first year students with a major in mathematics or physics. Our aim is to teach a *general control strategy* for problem solving. We emphasize that the solution process consists of five major stages, and that for each stage adequate heuristics can be selected. As the successful use of such heuristic strategies calls for good executive decisionmaking, we are convinced that higher–order thinking skills (metacognitive skills) play an important role in the problem–solving process. During the course we therefore try to teach our students some control strategies.

In this paper we describe the theoretical background, the content of the course and the method of instruction. We also discuss an evaluation of the course.

1. THEORETICAL FRAMEWORK

Since problem solving is an extremely complex activity, influenced by a large number of factors, we agree with Lester (1984) that a *holistic* approach to problem solving instruction is most appropriate. We no longer want to restrict ourselves to the teaching of heuristics, as we did before, see Grandsard (1988). Adopting a holistic view of problem solving justifies the use of *naturalistic* inquiry methods (see Lester 1985). We follow Glaser and Strauss (1967) in their argument that in the ill–defined field of problem solving it is sensible to derive theory from real classroom data, and that – in contrast to the so called scientific research pattern of pre–research → theory → testing the theory – this approach can be continued throughout the development of the theory.

Our work has been influenced quite a lot by Alan Schoenfeld's book on problem solving (1985). Our *control strategy* is based on the one he describes. The strategy is given in the form of a flowchart indicating the major stages of the problem solving process: analysis, planning, exploration, implementation and verification.

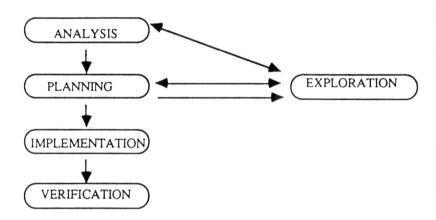

In the analysis stage you become familiar with the problem. Sometimes this results in an outline (plan) of the solution. However, most of the time an exploratory phase is necessary, a phase in which you examine knowledge, methods of proof, equivalent or more accessible problems. If you make progress you can either design a plan for the solution or re–enter the analysis phase. After a plan has been devised, it has to be implemented. In the verification phase the solution is checked. By reviewing the solution process, you can sometimes find alternative solutions or become aware of problem solving aspects that can help to improve your problem–solving capacities.

The flowchart is not be be interpreted as a program that the students are supposed to implement automatically. It is meant as a guide to use when a student does not know what to do next. So if a student knows what to do, he just does it. If he does not know what to do, the strategy suggests which heuristics might be appropriate, and in what order. If the student makes substantial progress, and reduces the original problem to a simpler or more manageable one, then the process starts over again.

2.   DESCRIPTION OF THE COURSE

Our problem–solving course is part of an interdisciplinary research project on specific and general thinking structures, with funding from the

Belgian FKFO. The course has been taught during the last two years to small classes of first year students (12 to 20 students) with a major in mathematics or physics. The 15 two-hour sessions are integrated in the normal curriculum (replacing 30 hours of scheduled exercise sessions). A pre-test and post-test are organised for research purposes only – to evaluate the student performance and the teaching strategy. Grading of the students is based on their activity in class and on homework.

In the course we work on a number of problems, such as the ones listed below, with the students. This is done either in class, or in small groups consisting of 2 or 3 students. We explain the role of the different stages in the solution of the problems, and the use of the flowchart as a guide when no ideas are available. Discussions are encouraged. We focus them on control decisions within the problem-solving process: selecting the right strategy, recovering from an inappropriate approach, monitoring and overseeing the problem-solving process.

Examples of problems
===

1. Find the smallest positive integer with 15 divisors. Are there smaller numbers with more divisors?

2. Let A be a set containing 17 elements. How many subsets of A have an even number of elements?

3. Let $B = \{1,4,7,10,...,100\}$ be the set of elements of a finite arithmetic sequence. Let A be a subset of B containing 20 elements. Show that A contains 2 elements whose sum is 104.

4. At a party 6 people are present. Show that one of the following situations arises:
   - either there are 3 persons who know each other mutually (in pairs)
   - or there are 3 persons who are total strangers from each other (in pairs).

5. Place the numbers from 1 to 10 on a circle in an arbitrary order. Show that there are 3 consecutive numbers whose sum is at least 17.

6. Find the product

$$\left[1 - \frac{1}{4}\right]\left[1 - \frac{1}{9}\right]\left[1 - \frac{1}{16}\right] \cdots \left[1 - \frac{1}{n^2}\right]$$

7.   Show that

$$\forall \ n \ \epsilon \ N_o \ : \ n! \ \leqslant \ \left[\frac{n+1}{2}\right]^n$$

8.   Given a circle, a line through its centre and a point outside the circle and not on the line, how can you construct (using a ruler only) the line through the given point perpendicular to the given line?

9.   Find all differentiable functions $f: \mathbb{R} \longrightarrow \mathbb{R}$ satisfying

$$f(x+y) \ = \ f(x) \ + \ f(y) \ + \ 2xy \ \text{for all} \ x,y \ \epsilon \ \mathbb{R}$$

10.   Find all continuous functions $f: \mathbb{R} \longrightarrow \mathbb{R}$ such that $f \cdot f = f$.

A great deal of attention is given to observing individual and small group problem solving. In the future we hope to investigate the role of metacognition in the problem–solving process and to incorporate metacognitive components in our problem–solving model.

When we discuss the different stages in the problem–solving process we give the students a handout with a detailed description of the strategy. This handout can be used whenever necessary. We hope that, after intensive use of the strategy during the problem–solving course, the students will start to use at least some of the heuristics and control strategies in other circumstances.

The problems we solve in our course are not mere exercises. Their nonstandard nature ensures that the students are not able to solve them simply by recalling and applying a familiar pattern or technique.

No problem can be solved without some basic knowledge or technique. To minimise the effect of a lack of knowledge or misconceptions on the problem solving activity, we choose simple problems, in the sense that they do not require sophisticated mathematics. Most of the problems are accessible to high school students in their last 2 years.

The problems are chosen in such a way that the following skills can be practised and emphasised: pattern recognition (and induction), choice of an efficient representation, translation into a suitable language or model, reformulation, regrouping of the data, reduction to a simpler or more accessible problem, and so on.

The role of the teacher in our problem–solving course is quite different from the usual one. In most classes the teacher explains, putting the students in a passive, imitative role. In our problem–solving class the students have to explain and to manage, whereas the teachers have a counselling role. They give guidance to groups pursuing different lines of thought and lead discussions if necessary. The hints that are given to the students are of heuristic nature and do not concern the content of the problem.   The guidance is oriented towards control: evaluation of ideas, decisionmaking, monitoring the problem–solving process.

3.   EVALUATION

After a first experimental version of our course during the previous academic year, we thought it important to evaluate the second modified version.   We wanted to investigate to what extent our goal, teaching a problem–solving strategy, was attained.   We have done this by comparing the problem–solving performance of the students who participated in the course to the performance of students who did not participate.

Measuring problem–solving abilities of students is mostly done by measuring the percentage of correct answers on tests.   We think that, in view of the complex skills that underlie problem solving, this is a very limited way of assessment.   The different stages in the problem–solving process demand measuring the ability at each step.   We therefore prefer a qualitative evaluation, with emphasis on the evalution of the problem–solving process, and not on the end product, (Malone, Douglas, Kissane and Mortlock 1980, Schoen and Oehmke 1980, Schoenfield 1982).

For our evaluation we have constructed a pre–test and a post–test, each consisting of five problems.   The tests are parallel, in the sense that corresponding problems ask for the same skills or heuristic strategies – analysis of special cases, translation of a word problem into equations, and suchlike.   Most of the problems can be solved in a variety of ways. The students had 20 minutes to work on each problem.

Two groups of students participated in the experiment.   The experimental group (11 students who major in mathematics) to whom the course was given, and the control group (16 students who major in physics or computer science).

For each problem we analysed and graded the following aspects: understanding (10% of the total score), analysis (40%), evolution of the problem–solving process and systematic method in the different phases (40%), implementation (10%).   We restricted ourselves to these aspects because we think that they are precisely the ones that can be observed in written protocols.

We also gave a positive correction to the score if we observed positive metacognitive attitudes, or a negative correction in the case of negative metacognitive attitudes.

The following histograms show the results of both groups for the pre– and post–test.   The scores are average scores over all students and all problems, the maximum score being 10.   We show a block diagram for the total score, for the score analysis, and for the score on the evolution of the problem–solving process.

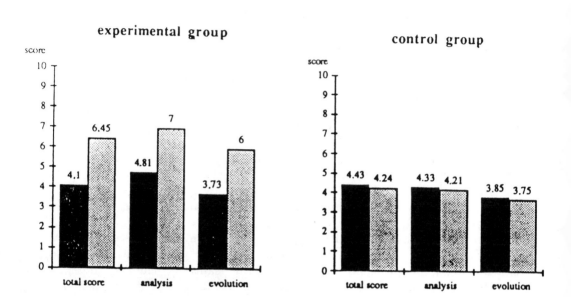

In view of the almost identical results for both groups on the pre–test, we can consider the groups to be comparable. However, the control group does not show any improvement from pre–test to post–test, whereas for the exerimental group there is a dramatic improvement. So there is clear evidence that the students learned something in our problem–solving course. One can of course discuss the nature of what they learned. Is the improvement due to improved decisionmaking, or is it a consequence of more problem–solving experience?

During the pre– and post–test the students had to answer, after each problem, a series of control questions concerning their problem-solving behaviour and their familiarity with the problem. Although the test problems were independent of the problems we had used in the course, the experimental group reported, at the time of the post–test, signs of recognition, at the level of the underlying stuctures, of the problems.

REFERENCES

Glaser, B. G. and Strauss, A. L. (1967). *The Discovery of Grounded Theory*. Aldine.

Grandsard, F. (1988).   A Computer-Assisted Course on Problem Solving and Methods of Proof.   *Problem Solving − A World View*, Proceedings of the Problem Solving Theme Group at ICME 5, Shell Centre, 274−278.

Grandsard, F. (1987).   *Problem Solving in de Wiskunde.*   Internal ISSAD report (in Dutch).

Lester, F. K. (1985).   Methodological Considerations in Research on Mathematical Problem Solving Instructions.   In Silver, E. A. *Teaching and Learning Mathematical Problem Solving: Multiple research perspectives.* Hillsdale: Erlbaum.

Malone, J. A., Douglas, G. A., Kissane, B. V. and Mortlock, R. S. (1980).   Measuring Problem Solving Ability.   In Krulik, S. and Reys, R. E. *Problem Solving in School Mathematics.*   National Council of Teachers of Mathematics.

Polya, G. (1971).   *How To Solve It.* Princeton, New Jersey: Princeton University Press.

Schatteman, A. (1987).   *Verslag van een cursus problem solving*, winter 1987.   Internal ISSAD−report (in Dutch).

Schoen, H. L. and Oehmke, T. (1980).   A New Approach to the Measurement of Problem Solving Skills.   In Krulik, S. and Reys, R. E. *Problem Solving in School Mathematics.*   National Council of Teachers of Mathematics.

Schoenfeld, A. H. (1982).   Measures of Problem Solving Performance and of Problem Solving Instruction.   *Journal for Research in Mathematics Education, 13(1)*, 31−49.

Schoenfeld, A. H. (1983).   Beyond the Purely Cognitive: Belief systems, social cognitions and metacognitions as driving forces in intellectual performance.   *Cognitive Science, 7*, 329−363.

Schoenfeld, A. H. (1983).   *Problem Solving in the Mathematics Curriculum.   A report, recommendations, and an annotated bibliography.   MAA Notes no 1.* The Mathematical Association of America.

Schoenfeld, A. H. (1985).   *Mathematical Problem Solving.* Orlando New York: Academic Press.

CHAPTER 23

Using Problem Solving Modes of Teaching to Promote Mathematical Thinking in Jamaican Secondary School Students

I. Isaacs
Mona University, Jamaica

ABSTRACT
A number of short—term and long—term intervention studies were conducted by students and staff of the University of the West Indies (Mona) during the period 1977 to 1985 in the upper grades of Jamaican High Schools to promote mathematical thinking. Various problem—solving modes of teaching were used such as group approaches to problem solving, direct modes of teaching and indirect modes of teaching problem—solving skills. At the end of the short—term studies there were measurable gains in the problem—solving skills of most of the students. In the studies extending over one or two years, gains observed during the life of the projects were not reflected in the students' performance on public examinations. In particular, students showed gains in their ability to conjecture, specialise and generalise, but not in their ability to reason deductively in general and abstract contexts.

1. INTRODUCTION

Most studies of Jamaican children's mathematical thinking have been done by graduate students at the Mona Campus of the University of the West Indies. Many of these trainee teachers were initially dubious of the feasibility of training students to think mathematically.

To change this perception, one needed to prepare teachers who would foster and promote high—level thinking in their mathematics classrooms. The mathematics staff in the Department of Education at

Mona directed their teacher preparation programmes to those teaching strategies which would enhance the development of mathematical thinking in secondary students, strategies such as inquiry learning, investigations, and problem solving. They also tried to influence their colleagues in the Jamaican Teachers' Colleges to develop a curriculum for their secondary teachers which would expose their student teachers to more divergent forms of learning and teaching mathematics.

Initially students on the diploma courses (pre-service and in-service) were dubious about the possible effectiveness of teaching the prescribed syllabus using problem-solving styles. Many of them readily accepted that other divergent styles of teaching, such as guided-discovery or laboratory, would be easier to manage and be more acceptable to the secondary students and their regular class teachers. On probing, it was admitted that many of the student-teachers felt insecure about problem solving itself. They admitted that they had got through most of their mathematics courses by memorising and algorithmic thinking, hence they were reluctant to teach what they had not practised. Those who could not be convinced that the diploma year was an opportunity to explore really new approaches to teaching were directed into trying the more straightforward techniques for developing divergent thinking in their pupils. With the introduction of the Caribbean Examinations Council mathematics examinations in 1979, the problem-solving component of the diploma program received an added boost. In this new examination, a candidate's performance would be reported on in three profiles: Knowledge, Comprehension and Reasoning. The last of these profiles included the ability to solve novel problems.

2.   STUDIES   DESIGNED   TO   PROMOTE   MATHEMATICAL   THINKING

Most of the intervention type studies done at the Mona Campus of the University of the West Indies were implemented in senior classes in high schools (McClean 1977, Buddo 1982, Carrington 1982). All the teachers who worked with senior students tried to impress on their project classes that the heuristics taught or developed were generally applicable to quantitative type problems. In one case (McClean 1977) the students were taught to use the heuristics with novel problems, both in mathematical and non-mathematical contexts.

McClean's study also differs from the other studies in that he used de Bono's (1972) guidelines for developing divergent thinking rather than the traditional set of heuristics usually attributed to Polya (1957). However, his class focused on those heuristics which are particularly apt for solving quantitative and logical problems, namely
(a)   concentrating on one aspect of a problem,
(b)   solving similar and related problems,
(c)   consideration of partial solutions,
(d)   working backwards.

This study was of short duration and was not followed up by later testing for the retention of the thinking skills taught.   McClean reported that the students improved in their skills immediately after the teaching episode, but it did not seem to have any carry over effect on their performance on the Cambridge GCE A-level mathematics examination.

The other studies used the guidelines given by Polya (1957) in How to Solve It to determine the heuristics to be taught.   These heuristics were made to arise out of the context of the problems posed to the students.   The results from these and other problem-solving studies carried out at Mona indicate that, over a short period of time

(a)   some problem-solving skills can be developed in specific topic areas such as trigonometry, calculus, and analytical geometry at the Advanced level, and solving equations and inequalities at the General and Basic levels;

(b)   the grade level at which problem-solving skills seem to most rapidly develop is at grade 10;

(c)   the teaching approach which seems to be most effective is the one which uses selected topics which are appropriate for illustrating heuristics, such as
      (i)   systematic trial and error,
      (ii)   drawing diagrams (graphs) and making tables,
      (iii)   examining simpler or special cases,
      (iv)   considering different approaches,
      (v)   working backwards.
Carrington (1982) introduced the double-angle science formula to a class of grade 12 girls by having them consider the following problem.

If sin 30° = 1/2 then determine sin 15° without using tables.

In trying to solve this problem the girls were made aware of the various problem-solving strategies they were using to come to grips with the problem.   Strategies such as drawing a diagram, relating the data and the problem to previously acquired knowledge, and considering different approaches to solving a problem were highlighted by the teacher.   Some of the students conjectured that the bisector of an angle of a triangle divides the opposite side of the triangle in a ratio corresponding to the length of the arms of the angle.   This conjecture led the class into trying to prove this theorem which was new to all of them.   The attempts to prove this theorem provided yet more opportunities to practise problem-solving strategies.   Carrington followed these initial classes with the derivation of the general formulae for the double angle for the sine and cosine ratios.   Later work on the trigonometric identities gave him, and the class, many opportunities to use the strategies of working backwards and working from both ends to show that two expressions were identical.

Buddo (1982), teaching an average grade 10 class in a coeducational high school how to solve inequalities, found that she was able to arouse considerable interest in this topic by initially posing verbal problems such as the following.

> A class consists of boys and girls.   Each boy contributes twice as much as each girl towards a class present for their class teacher. In all they collect $43.   What is the smallest number of girls present in the class?

In the class discussion which ensued students pointed out that they were not told how large the class was, nor the size of each contribution, nor the relative size of number of boys to girls.   Buddo suggested that they break up into subgroups to discuss these points, and come to some sort of consensus on these issues.   The class was then divided into friendship groups which, with the help of the teacher, engaged in heated and animated discussion with the teacher acting as a moderator and referee.   Eventually a number of possible solutions were explored using different parameters.   The teacher summarised their activity by focusing the students' attention on the stages and strategies they had used to solve the problem.

Features   of   problem–solving   behaviour   (such   as   clearly comprehending the conditions stated and implied in a problem, testing different cases, working back from the answer, and checking that the solution fits the initial conditions of the problem) were listed on the chalkboard by the teacher.   In the following classes the teacher moved from similar unstructured problems to more structured problems from which students could derive linear inequalities.   These they could solve either by systematic trial and error or by graphing.   At the end of the project, Buddo found that the students had developed a strong liking for this topic and a positive attitude to problem solving in mathematics. She attributed this partly to the group approach used and partly to the non–threatening environment in which the problems were posed and solved.

3.   THE EXTENDED STUDIES

Studies of a longer duration to improve mathematical thinking have been carried out by Isaacs (1978, 1983, 1986) and Williams (1986).   Isaacs, over the period 1977 to 1985, carried out a number of extended intervention type studies lasting one or two years at a time to develop directly the mathematical–thinking skills of senior secondary students in Jamaican High Schools.   He conjectured that students' problem–solving skills would only become part of the students' permanent behaviour pattern if they were practised over long periods in the context of their regular mathematics programme.

In the first study (1978) Isaacs taught a class of grade 12 mathematics students over the period of one academic year a selection of the problem—solving heuristics of Polya (1957), using selected topics from the Cambridge A—level mathematics syllabus to illustrate them.   The aims of the teaching project were

(a) to improve the problem—solving abilities of first year sixth form (grade 12) mathematics students by teaching some of the heuristics for problem solving in a normal classroom setting,
(b) to develop flexibility of approach in the students' thinking,
(c) to develop the students' ability to think reflectively.

The sample consisted of twenty—one students (17 males, 4 females) in their first year (grade 12) at a Community College in the corporate area of Kingston and St Andrew.   These students planned to take the traditional Cambridge A—level papers in mathematics after two years in the college.   Sixteen of them had obtained grades of A or B in the Cambridge O—level mathematics examination.   Eighteen of them had done the traditional O—level papers and three had done the modern syllabus.

The class was taught using an *indirect problem—solving teaching style*.   This approach requires the teacher to select topics from the prescribed syllabus which can be introduced by illustrative examples which

(a) can be solved in more than one way,
(b) embody in themselves some of the salient characteristics of the general problem,
(c) will encourage the students to inquire how the problem can be solved in general.

Topics such as the solution of algebraic equations of degree higher than two, the derivation of the double—angle identity for the trigonometric ratios, the development of the derivatives of x^n and $\sin x$ and the problems in analytical geometry, provided a fruitful source of exercises for practising the heuristics of problem solving.

The topics, wherever possible, were taught in a manner different from that used in the students' textbooks.   If the approach were similar, illustrative material different from that in the textbook was used.   For example, to introduce the solution of algebraic equations using the remainder theorem, the following problem was posed.

(a) A rectangle has an area of 240 cm² and perimeter of 62 cm. What are its dimensions?

(b) A triangle has an area of 240 cm² and a perimeter of 62 cm. What are the lengths of the sides if it is
     (i)   an isosceles triangle,
     (ii)   an equilateral triangle,
     (iii)   a scalene triangle?

In trying to solve part (b) of this problem students derived the equation

$$h^3 - 31^2h + (31)(480) = 0$$

The students were encouraged to solve this equation in any way they could. They tried to solve it by using the quadratic formula, graphing, trial and error (with the assistance of a pocket calculator), and the remainder theorem.

The teaching/learning situations were organised around questions or problems raised by the students or the teacher. The teacher emphasised conflicting, difficult or puzzling (seemingly paradoxical) aspects of the topic. For example, in the problem posed above, students were disturbed to find that there were no solutions to some parts of the question, or that the solutions found were not applicable to the problem. In the project, questions were asked which were not readily answered by recalling definitions, theorems, or illustrative examples previously shown by the teacher or to be found in the textbooks. To develop flexibility in thinking, the teacher encouraged the students to make multiple suggestions for tackling problems, and to show on the chalkboard two or more ways of solving a given problem, with the teacher adding alternative solutions where necessary, and identifying and explicitly stating the different heuristics used in the solution of the problems. In the case of the example cited above, at the end of the sessions the teacher drew to the attention of the students the number of different strategies they had used to attack the problem. He listed them on the chalkboard and emphasised these strategies as the important tools of the successful problem solver.

When a substantial body of heuristics had been discovered students were encouraged to consult this list and choose likely ones before tackling a given problem. When the problem was solved, or a major blockage occurred, the teacher would lead a discussion on the feasibility of the heuristics selected and the possibility of using alternative ones in the given situation. Finally, the teacher modelled in front of the class the problem–solving processes *he* used, by thinking aloud and exposing to the students *his* ways of approaching novel problems posed by the students, which were extracted from their textbooks and past examination papers.

Problem–solving tests were administered at regular intervals during the year. Four of the students showed discernable growth in their problem–solving abilities during the year. The rest of the class showed no consistent changes in their problem–solving behaviours. Under formal examination conditions nearly all the students reverted to the rigid types of thinking shown at the start of the project. Their performance was not significantly different from a comparison group in another school who had followed a more conventional approach to the A–level course.

This study, together with a follow–up study (Isaacs 1983), led the investigator to conclude that, to develop flexibility and willingness to try new approaches, one has to start much earlier in the school programme. It seems that students who have succeeded at school mathematics after 11 years of schooling have developed a perception of mathematics and what is good mathematics teaching that is antithetical to the higher levels of thinking required for doing mathematics effectively beyond the elementary stages of the subject.   Furthermore,   the   types   of mathematical thinking at which these students excel are rapidly being replaced by modern electronic computing machines.   It is also possible that another approach to teaching problem solving might be more effective with students coming from a background of rote learning and drill and practice of algorithms.   For such students a more direct approach,   such   as   that   developed   by   Charles   (1982),   might   be   more effective.

Isaacs' third study   (1986, 1987a, 1987b) aimed to investigate again in natural classroom settings, which, if either, of two problem–solving teaching styles was likely to promote mathematical thinking and what was their effect on mathematical achievement in public examinations.   Two grade 10–11 classes in an urban Jamaican High School were taught over a period of five terms in two problem–solving styles: an Explicit Style (Charles 1982, Charles and Lester 1982) and the Implicit Style (Isaacs 1978) developed by the investigator (the style described above in the first study).   The Explicit Style of teaching problem solving requires a teacher to

(a)   compile sets of problems (mainly novel problems) which exemplify certain specific or general heuristics,

(b)   sequence the teaching so that problems illustrating the simplest heuristics are introduced first, then follow these by sets of problems illustrating the next level of heuristics, and so on until the final set covers all of the heuristics dealt with in the teaching programme,

(c)   encourage students to compile a list of strategies (heuristics) in their notebooks and display the list frequently during problem–solving sessions,

(d)   apply the problem–solving strategies whenever possible to problems and exercises in the mainstream course.

Two intact science–oriented classes from the fourth form (grade 10) were assigned to the investigator for this project.   The investigator randomly assigned the classes to the two teaching styles.   Both classes were found, at the start of the project, to be comparable on measures of achievement, problem solving and logical reasoning.   During the first phase of the project, students following the Explicit Style of teaching spent one period per week working on and discussing novel problems which illustrate some of the heuristics of problem solving.   The heuristics (called strategies in the class) explicitly taught were

(a)   understanding the problem,

(b)   drawing a diagram, or making a table, or writing an equation,

(c)   looking for a pattern,

(d)   systematic trial and error, or drawing a graph,

(e) trying a simpler case or cases,

(f) using deductive arguments,

(g) checking the solution.

In the other class following the Implicit Style of teaching, problems were used to introduce new topics, or introduced by the investigator from their textbook or from student queries. A conscious effort was made not to devote more than one period per week to problem-solving tasks in either of the classes. The remaining five periods per week were devoted to teaching topics from the mathematics syllabus of the Caribbean Examinations Council in a guided discovery style based on the students' textbook (Mitchelmore, Raynor and Isaacs 1976). Topics covered were Sets and Logic, Relations and Functions, Introductory Number Theory, Transformation Geometry, Vectors, Statistics and Probability, Consumer Arithmetic and Computation.

Three problem-solving tests were administered during the first year of the project. For each of these tests three students from each class, matched for mathematical thinking ability and achievement, were audio taped so that detailed studies could be made of their thinking processes. They were also interviewed immediately after completing each test for their views and comments on the problems, and how they had tackled them. The other students were encouraged to show all their working and hand in all rough work done during the test. At the beginning of the project most students in both classes could make some progress in solving problems which required examining simpler cases, recognising simple numerical patterns, and making direct inferences from data given. However, they were unable to recognise invalid patterns of argument. During the year it was observed that their use of some heuristics increased (drawing a diagram, looking for patterns, examining simpler cases, and checking solutions). They were reluctant to record deductive arguments and justifications. Even the more able students, who were quite willing to say why their results were reasonable, were unwilling to put it down in writing.

In June 1984 the students sat the CXC's Basic Proficiency papers in mathematics (Caribbean Examinations Council 1984), the Mona Reasoning Test – Form A (Brandon 1984) and the investigator's End of Year problem-solving test (Isaacs 1986). On all three instruments the two classes performed at about the same level.

The performance of the two classes on the Reasoning Profile of the CXC Basic papers was also compared with the performance of the population which standardised these papers, and was found to be about one standard deviation above the population's mean score (Isaacs 1987a). However, as the two classes were about 1.3 standard deviations above the population on the whole examination, it was not reasonable to infer that their training in problem solving had made any significant change in their ability to use these skills under examination conditions.

Phase 2 of the project spanned the period September 1984 to May 1985. Until Februrary 1985, both classes continued to have problem

sessions devoted to novel or hard problems related to the mainstream course.   All the problems examined during the first two terms of the 1984–85 academic year were taken from topics on the CXC's General Proficiency syllabus (1981).   Problems related to
(a)   finding angles in circles,
(b)   proving theorems in geometry (vector, synthetic and analytical),
(c)   transformation of plane figures by matrices,
(d)   vectors applied to physical situations,
(e)   trigonometry applied to physical situations,
(f)   solving algebraic equations arising from physical solutions,
were the main source of stimulus material for the problem–solving sessions.   At the end of February the Final Problem–Solving Test (Isaacs 1986) was administered.   For the remainder of the teaching sessions, until May, the problem–solving sessions were used to do questions from previous CXC and GCE papers at the General Proficiency level.   In June 1985, 58 of the students (29 in each class) sat the CXC's General Proficiency Papers in Mathematics (Caribbean Examinations Council 1985).

On the investigator's Final Problem–Solving Test, the two classes did much better than they did at the start of the project and at the end of phase one of the project.   However, it is difficult to compare statistically their actual mean scores, as the three tests were different and were only comparable in the types of processes tested in the problems.   When the classes are compared with the population who took the General Proficiency papers they scored about 0.9 of a standard deviation above the mean of the population on the Reasoning Profile as well as on the traditional profiles of Computation and Comprehension.   A post hoc analysis of the data (Isaacs 1987b) revealed that the 'average' students in the class taught in an implicit problem–solving style did as well as their 'above average' classmates on the traditional profiles of the CXC papers. This finding suggests that the 'average' students in the implicit class learned to apply the problem–solving heuristics taught to the routine exercises of the CXC mathematics programme.   If this is so, then this study suggests that 'average' students taught in an Implicit Style are more likely to apply problem–solving skills to the mainstream course than those taught in an Explicit Style.

One area in which there was no discernable improvement over the two years was in the students' ability to detect false arguments based on indirect implications.   Even the above–average students, who could classify implications as direct and indirect and were able to state that the converse and inverse of a true implication is not necessarily true, were frequently misled if the statements to be examined were embedded in a socially realistic context.   Their exposure to the Reasoning and Logic unit of the syllabus did not seem to help them cope with these problems.

Williams (1986) attempted to improve the formal and general reasoning skills of a sixth form class at a rural High School in Jamaica by writing a programmed unit on Reasoning and Logic incorporating mathematical and real-world examples to illustrate the principles and laws of mathematical logic. Over the period of an academic year the students worked on the unit on their own, with individual assistance from the investigator. During the course of the project she studied in depth the background, interests and academic performance of three of the students, with a view to seeing how non-cognitive variables influenced their performance on mathematical tasks. At the end of the project the students were tested on their reasoning skills as well as their achievement in A-level mathematics. Williams found that they had improved in their ability to reason, but that this ability did not correlate with their performance on the A-level mathematics papers. From a content analysis she concluded that, to do well on the A-level papers, this ability was not required.

4.   DISCUSSION

The intervention studies done by students and staff at the University of the West Indies (Mona) indicate that it is possible to teach older students some of the general mathematical thinking processes, particularly the processes of specialising, conjecturing and generalising (Burton 1984). However, it is not clear that problem-solving styles of teaching are any more effective in developing these general traits than say a guided-discovery or laboratory approach to teaching mathematics. However, teaching mathematics in these ways is making no more contribution to the curriculum than teaching science by inquiry modes. In both approaches we are developing inductive thinking, only the objects of study are different – in the case of science, physical and biological objects, and in the case of mathematics, numerical, symbolic and spatial objects.

Mathematics differs from science in the method of justifying generalisations. What the students have to come to realise in mathematics is that they have to produce an argument that will first convince themselves, then others, that their generalisation must be true. In so doing they have to begin to think critically about their conjectures, processes of arriving at the solution, and the solution itself. It is this unique feature that problem-solving modes of teaching contribute to the development of mathematical thinking.

REFERENCES

Brandon, E. O. (1984). The Mona Reasoning Test, Form A. Faculty of Education, University of the West Indies, Mona.

Buddo, C. (1982).   A Problem Solving Approach to Teaching Linear Inequalities to Grade 10 Students at a High School.   Unpublished Dip Ed study; U.W.I., Mona.

Burton, L. (1984).   Mathematical Thinking: The struggle for meaning. *Journal for Research in Mathematics Education, 15(1),* 35–49.

Caribbean Examinations Council (1981).   *Mathematics Syllabus.*   The Registrar, St Michael, Barbados.

Caribbean Examinations Council (1984).   Secondary Education Certificate Examination, Mathematics Basic Proficiency, Papers 1 and 2.   The Registrar, St Michael, Barbados.

Caribbean Examinations Council (1984).   Secondary Education Certificate Examination, Mathematics General Proficiency, Papers 1 and 2.   The Registrar, St Michael, Barbados.

Carrington, M. G. (1982).   The Development of Problem Solving Strategies in a Group of Sixth Formers.   Unpublished Dip Ed study; U.W.I., Mona.

Charles, R. I. (1982).   An Instructional System for Mathematical Problem Solving.   In Rachlin, S. L. (ed).   *Problem Solving in the Mathematics Classroom,*   17–32; Mathematics Council of the Alberta Teachers' Association, Calgery, Alberta.

Charles, R. I. and Lester, F. K. (1982).   *Problem Solving: What, Why and How.* Dale Seymour; Palo Alto, Ca.

de Bono, E. (1972). *PO: Beyond Yes and No.*   Penguin Books; Harmondsworth, England.

Fuller, L. (1983).   A Problem Solving Approach to the Calculus. Unpublished Dip Ed Study; U.W.I., Mona.

Isaacs, I. (1978).   Teaching Students in Sixth Form to Use Some of the Heuristics for Solving Problems in Mathematics.   Mimeo report; School of Education, University of the West Indies, Mona.

Isaacs, I. (1983).   Teaching Problem Solving in a Sixth Form College Within the Confines of a Prescribed Syllabus.   In Zweng, M., Green, T., Kilpatrick, J., Pollak, H. and Suydam, M. (eds) *Proceedings of the Fourth International Congress on Mathematical Education,* 282–283. Birkhäuser; Boston, Ma.

Isaacs, I. (1986).   Nurturing Mathematical Thinking.   Mimeo Report; Faculty of Education, University of the West Indies, Mona.

Isaacs, I. (1987a). Using Problem Solving Teaching Styles to Prepare Candidates for the CXC Basic Proficiency Examinations. *Educational Studies in Mathematics,* **18**, 177–1990.

Isaacs, I. (1987b).   The Effects of Extended Teaching in Problem Solving Styles on the Mathematical Achievement of Secondary School Students in Public Examinations.   In Sit–Tui Ong (ed) *Proceedings of Fourth Southeast Asian Conference on Mathematical Education*, 162–167; Institute of Education, Singapore.

McClean, B. (1977).   Can Problem Solving in Mathematics be taught to 6th Form Students?   Unpublished Dip Ed study; U.W.I., Mona.

Mitchelmore, M. C., Raynor, B. and Isaacs, I. (eds) (1976).   *JSP Caribbean Maths*, Textbook 4.   Longman Caribbean; London.

Polya, G. (1957).   *How to Solve It* (2nd ed).   Princeton University Press; Princeton, NJ.

Schoenfeld, A. H. (1976)   Can Heuristics be Taught?   Unpublished report; SESAME Project. Physics Department, University of California, Berkeley.

Williams, N. (1986).   The Formal and Logical Reasoning Ability of Some Sixth Form Students.   Unpublished MA thesis; U.W.I., Mona.

CHAPTER 24

Trends and Barriers to Applications and Modelling in Mathematics Curricula

J. de Lange
Research Group on Maths Education, OW & OC, Utrecht University, The Netherlands

ABSTRACT

In a recent article, Niss presents a survey of post–war developments and present states and trends in the role of applications and modelling in mathematics curricula at all levels (Niss 1987).

In this article we propose to discuss the trends and barriers that Niss has noticed in the light of the actual situation in the Netherlands.

1.   TRENDS

A general trend seems to be the fact that, in countries with *central* curriculum authorities, the inclusion in post–elementary curricula of proper applications, models and model building on a large scale has *not* been brought very far.   In countries in which the curriculum authority is *decentralised*, examples of curricula in which non–trival application and modelling activities are implemented are most abundant.

Niss relates this fact to the introduction of new math.   It seems to him that in countries where the new math reforms were implemented early, and hence took its most pure shapes, the inclusion of applications and modelling in the curricula has been brought furthest.

Let us first state that we agree with this general picture.   More interesting, however, is the situation in the Netherlands.   There we see applications and modelling in the mathematics curriculum (since 1985) at the request of a Committee installed by the Central Government, after experiments carried out by a National Research Group and introduced by the Government for all schools in 1985 (de Lange 1987).

On the other hand, new math reforms never quite made it in the Netherlands.   Most of the wave that originated at the Royaumont Conference in 1959 (Fehr et al 1961) was stopped by a dike that was built by Freudenthal and his Wiskobas–IOWO colleagues.   Only in secondary education did some of the modern math get a foothold.

The reasons behind the introduction of applications and modelling in the Netherlands are clear.   Even before some of the new math entered the curricula, it was felt that formal and abstract mathematics education did not serve the purpose of educating the non–specialist.   It was a recognition of the increased mathematisation of many other fields than physics: mathematics has be to useful.

An important period in the discussion of the role of applications and modelling was in 1976 when a conference was held in Utrecht entitled How to Teach Mathematics so as to be Useful.

Here the view was broadened from the world of physics to the real world in general, and the usefulness of mathematics was stressed.   In this conference we noticed pleas for using this real world to teach mathematising, while teaching math concepts along the way.

The first argument – the usefulness of mathematics – is, in the Netherlands, strongly intertwined with a second one.   This is that applications and problem solving should not be reserved for consideration only *after* learning has occurred; they can and should be used as a context within which the learning of mathematical ideas takes place (Lesh et al 1983).   Or, as a number of American mathematicians stated in their famous memorandum in 1962:

"To know mathematics means to be able to do mathematics: to use mathematical language with some fluency, to do problems, to criticise arguments, to find proofs and, what may be the most important activity, to recognise a mathematical concept in, or to extract it from, a given concrete situation.

Extracting the appropriate concept from a concrete situation, generalisation from observed cases, inductive arguments, arguments by analogy, and intuitive grounds for an emerging conjecture are mathematical modes of thinking." (Ahlfors et al 1962)

It is this 'extracting the appropriate concept from a concrete situation' that forms the backbone of realistic mathematics education, as proposed by the research group OW & OC.   One of the projects carried out by this Research Group was the so called Hewet Project. This was carried out between 1981 and 1987 and lead to a new applications–oriented curriculum for upper secondary math education. The experimental student materials developed during this project all share the idea of 'extracting the appropriate concept from a concrete situation'.

Depending on such factors as interaction between students and teacher, the social environment of the students, and the ability to formalise and to abstract, the students will sooner or later extract the mathematical concepts from the real world situation.   We refer to this phase as *conceptual* mathematisation.

After the development of the desired mathematical concepts, followed by a more strict and formal definition, we reach the phase of *applied* mathematisation. This brings us back to the usefulness of mathematics.

2.   BARRIERS

Niss poses the question 'Why is it so difficult in practice to ensure applications and modelling an appropriate position in mathematics curricula?'

For secondary education there seem to be three barriers, three keypoints.

1. Many *teachers* are not ready to accept a reduction of the purely mathematical syllabus in order to introduce application and modelling activities.
2. The second barrier lies with the *students*: many examples are associated with situations and problems which are no less academic and no less alien to students than are purely mathematical problems.
3. Application and modelling qualifications, especially in handling real problems, are difficult to *assess* and *test* by traditional evaluation tools.

During the experiments in the Netherlands we also focussed on these three keypoints. Our empirical research considered especially the problems of assessment and testing. However, we will first discuss teacher and student reactions to the introduction of the new curriculum.

2.1 Teachers

All reactions from teachers and students were from the period 1981-1985, that means the period in which the experiments took place as carried out by the Hewet team. First at two schools, later at ten, finally at another 40. We start with teacher reactions from the ten schools.

The teachers from the ten schools followed a teacher-training course given by the members of the Hewet team. This was a facility which had not been available to the teachers from the two schools. During this course the teachers were probed in order to learn about experiences in the pre-experimental tenth grade. Probes were taken in December 1982, March 1983 and May 1983. During the following years the contact between the Hewet team and the ten schools consisted of a number of meetings to discuss the experiences. Internal written reports exist of all meetings, from which we have extracted a general picture of the trend of experiences at these ten (twelve) schools. These meetings took place in October 1983, January 1984, May 1984, November 1984 and May 1985.

During the meetings we had the opportunity to classify the experiences. The first question at all meetings was always 'What are your experiences and your general impression?'   We have divided the answers given into three classes:

(a)   Teachers' difficulties, troubles etc were classified as '−'.
(b)   Teachers' mixed feelings, or neutral feelings, were classified as '0'.
(c)   In cases where teachers were positive or very positive we classified the experiences with '+'.

In this way we get the following table, resulting in the graph below it.

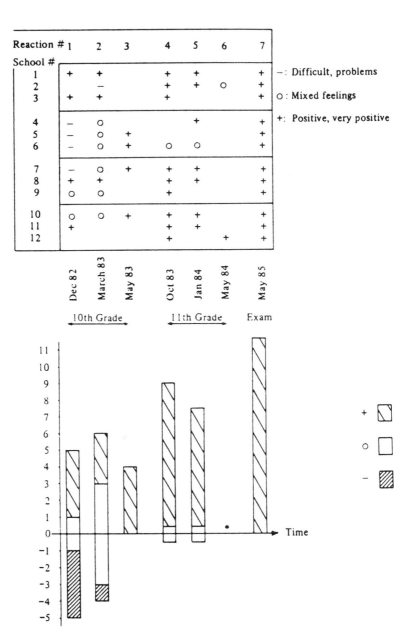

We have to be careful when interpreting these data because of the arbitrariness of the classification.   One conclusion, however, is clear: after an introductory period, where some doubts were cast on the new programme, there is a steady increase in the rating of the experiences. This positive trend becomes even stronger if we look at the ten schools (numbers 1–10), because of the fact that the two schools performed the new programme for the second or third time.

Next we proceed to the experiences from teachers of the 40 schools.   At the 40 schools stage the distance between the schools and the Hewet team was considerable.   The teacher–training course was not given by the members of the Hewet team but by professional teacher trainers.   On several occasions, however, the team was present at these courses, as were the teachers from the ten and two schools, to share classroom experiences.   This proved to be very useful.   At two opportunities, in February and October 1985, the team met with the majority of the teachers from the schools during two 24–hour conferences in order to exchange information and experiences.   We received reactions from 45 schools, out of a maximum of 50.

In a questionnaire we asked the teachers the question 'What has been your experience with Mathematics A so far?'   We did not use any five– or seven–point scale, nor did we point out special points of attention, in order not to influence the teachers.

The answers – of course – were very different in quality and quantity.   Long–winded ones were compressed, without loss of essentials.

We give a representative sample.

- We managed to carry on the whole programme without losing one student.
- Not only we, the teachers, are enthusiastic about the experimental materials, but the students as well, which may be illustrated by a comment of one of the students: 'The way we were working with math A the past two years filled me with real interest for mathematics'.
- My experiences are positive.   Mathematics A gives a new dimension to school mathematics.   What struck me is that some students were quite inventive and creative when interpreting results.   Both students and myself are increasingly enthusiastic and we work with pleasure.

Also in this case the reactions to Mathematics A were more positive than half a year before when the first probe was taken.   We have tried to tabulate the results of the latest probe available in order to get an overview.   For this purpose we initially classified the reactions in *four* classes: *negative, neutral, moderately positive* and *positive*.   However, none of the reactions was negative.   This leaves us with these classes:

Neutral	10 or 22%	based on 90% of schools
Moderately positive	13 or 29%	
Positive, Very positive	22 or 59%	

The table leads to the conclusion that experiences are better than moderately positive.   At the same time, however, we would bear in mind that all schools volunteered for these experiments, and are therefore by no means representative of the total population of 450 schools which introduced the new programme in the summer of 1985.   It *only* represents the experiences of the total group of experimental schools that worked with the *experimental student materials* developed by the Hewet team.

2.2 Students

- 'Compared to the maths we used to have I find this more difficult because you have to have more insight for this and that's something difficult for me.'

- 'I like Math A better than the Math I we had in the junior years. At least I can see what I'm doing now – what I'm calculating and to what purpose.   But on the other hand I find it difficult that there are so few formulas.   You have to really reason logically, you can't just blindly apply a formula.   And I often have trouble with that.'

- 'I find the new math a lot more pleasant; you see more clearly how mathematical formulae can be used in daily life.   In addition I think it's good that you're not just given a formula to use on a x–number of sums but that you have to derive a formula by means of contextual problems.   That way you understand a lot better how you've arrived at a formula.'

For more reactions we refer to de Lange (1989).

2.3 Assessment and tests

The Netherlands has nationwide central examinations.   The role of these examinations on the actual teaching and learning in the classroom has been discussed for a long time.   Van Hiele stated in 1957

"But soon, much too soon, the influence of the exam becomes noticeable, with the result that the exam makes it necessary to anticipate a long time in advance.   If at all, these preparations contribute to concept forming only in a negative sense."   (Van Hiele 1957)

These general problems became more clear when the new curriculum was introduced.   The usual, timed, written test is not appropriate to operationalise all goals of the new programme.   This is mainly because of the fact that the new programme stresses process skills like mathematisation, reflection, creativity, critical attitude, and mathematical thinking.

Research carried out among the teachers of the twelve experimental schools showed the following clearly.

1. It is very difficult (for teachers) to construct tests for math, especially because some goals cannot be operationalised appropriately by means of timed written tests.

2. It is very difficult (for teachers) to derive the goals of the new curriculum from the experimental student materials.

The conclusion we draw from our experiences coincides with the conclusion from the report Mathematics Counts. The assessment needs to take into account not only those aspects of learning which can be examined by means of timed written tests, but also those which need to be assessed in 'some other way' (Cockcroft et al 1982).

Given the results of our study among teachers of the 12 schools we developed assessment tests based on five principles:

1. to improve learning,

2. to demonstrate what students *know* rather than what they do not know,

3. to *operationalise* the goals of the course (particularly the higher order goals),

4. to accept competent judgement of scoring rather than objectivity,

5. only to use tests that can be readily carried out in school practice (de Lange 1987).

The last principle means that project–like tests were rejected as requiring too much investment in time. Assessment methods considered were traditional, restricted–time, written tasks, oral tasks, take–home tasks, two–stage tasks and essays. An innovative idea is the two–stage task: the first stage is a restricted–time written test of a more open kind, which is marked and returned with comments, for the student to take away and improve in their own time to obtain the second–stage mark.

Reviewing the different methods it was concluded that

The best possibilities are offered by a combination of restricted–time written tasks, take–home tasks (or combined in one two–stage task) and oral tasks which clearly measure different aspects of learning.

As a side effect of the new course, it was noted that:

It is an interesting outcome of the study that girls perform rather poorly on restricted–time written tests when compared to boys. This difference disappears almost completely in the case of alternative tasks that were part of this study.

3.   CONCLUSION

It will come as no surprise that the critical barrier to the introduction of applications and modelling in the mathematics curriculum is formed by the teacher.   We agree with Niss that there are teachers who consider the new course as a decline in the quality of mathematics education, which is not compensated for by the gains resulting from the application and modelling activities.   This is partly caused by the traditional grounding of teachers.

The problem is, in our opinion, mainly a matter of attitude.   There are many changes for the teacher:

- mathematics loses much of its structure,
- mathematics is not certainty,
- more answers are possible,
- students can have better solutions than teachers,
- more interaction,
- problems with testing,
- problems with basic skills.

However, it is also our experience that a good in-service teacher training course can readily change teachers' attitude.   The mistake made in the Netherlands is that not enough attention has been given to this aspect of the innovation.

The second barrier – the students – was not experienced as such in the Netherlands.   As our students' reactions showed, many of the students are relatively positive and, in fact, student participation has risen considerably because of the new curriculum.

Much attention is needed in developing good and relevant examples – this is clear.   However, in our opinion, this barrier seems to be much less critical than the other two.

Assessment and tests indeed form a critical barrier.   Although our study gave indications how to proceed to more appropriate tests, this does not mean that there is any hope of solving this problem in the foreseeable future.   Here we deal not only with old traditions, but also with political forces.   The battle in the Netherlands has just started – the first two national examinations were close to disaster, both from the point of view that no justice was done to the curriculum but also, and maybe even worse, that no justice was done to the students.

REFERENCES

Ahlfors, L. V. et al (1962).   On the Mathematics Curriculum of the High School.   *American Mathematical Monthly*, **3**.

Cockcroft, Sir W. et al (1982).   *Mathematics Counts*.   London, HMSO.

Fehr, H. F. (1961). *New Thinkings in School Mathematics.* Cambridge, Cambridge University Press.

Hiele, P. van. (1957). *Begrip en inzicht.* Purmerend, Muusse.

Lange, J. de, (1987). *Mathematics Insight and Meaning.* Utrecht, OW and OC.

Lange, J. de, (1989). The Teaching, Learning and Testing of a Mathematics for the Life and Social Sciences. In Blum, W. et al (ed) *Applications and Modelling in Learning and Teaching Mathematics.* Chichester, Horwood.

Lesh, R. et al (1983). *Acquisition of Mathematics Concepts and Process.* New York, Academic Press.

Niss, M. (1987). Applications and Modelling in the Mathematics Curriculum – State and Trends. *International Journal for Mathematical Education in Science and Technology,* **18(4)**, 487–505.

CHAPTER 25

Japan–US Collaboration in Cross–Cultural Research on Students' Problem–Solving Behaviours

T. Miwa
Tsukuba University, Japan.
J. P. Becker
Southern Illinois University at Carbondale, USA.

1.   INTRODUCTION
Japan–US collaboration in cross–cultural research on   students' problem–solving behaviours has begun in earnest.   This is an important development whose origin can be traced from the early 1970s, the first stage of which culminated in the US–Japan Seminar on Mathematical Problem Solving which was held at the East–West Centre in Honolulu, Hawaii, in July 1986.   The Proceedings of the seminar were edited by the authors, and are now available from Jerry Becker.

2.   PROMINENT IDEAS DISCUSSED AT THE SEMINAR
It is not possible to summarise, even briefly, all the papers/discussions in this short paper.   Instead, we have selected a few salient points from the discussions to report here (excerpted from Becker and Miwa 1987).

(a)   Japanese students found the problem–solving items on the Second IEA Study more routine than students in other countries. However, like American students, Japanese students view mathematics as very much static, and not dynamic and functional as it should be.   This finding has led to research on how to develop an open–minded and dynamic attitude towards mathematics and problem solving in Japan.
(b)   The impressive level of mastery of computation among Japanese students may be the result of a clear understanding of the decimal

numeration system and the high expectation of students by parents, teachers and society.

(c)   There is very little research information dealing with the fundamental issues of how students generate problems on their own, and how we transfer ownership of the problem from the teacher or curriculum developer to the students, so that it becomes the *student's* problem.

(d)   Though, in general, mathematics educators and psychologists are beginning to develop a common language, they tend to view mathematics problems differently – psychologists tend to view problems as an area in which to study human thinking, whereas mathematics educators tend to give more importance to the mathematics in a particular research setting.

(e)   There are different kinds and levels of understanding in mathematics and problem solving, and only an imprecise use of language among researchers, which leads to confusion in trying to deal with them. We know of no research source which delineates among these different kinds of understanding nor the relationship between them. To make progress in understanding problem–solving processes, we need to attack questions from as many different perspectives as possible.

(f)   Though problem solving is an important classroom objective in both countries, there is little evidence that teachers actually incorporate it as a *major* objective in their classrooms. Is this a consequence of their pre–service education, in which teachers have far *too* little experience themselves in problem solving? In Japan, problem, at least at the secondary level, means *problems on the entrance examinations* which, though original and not routine, may not be related to the real world.

Part of the difficulty in this area is finding problems whose situations are interesting to students and whose difficulty level is suitable for students. Another part of the problem is that standardised tests in mathematics commonly stress computation and, so long as this is the case, students have little opportunity for real understanding and development of their reasoning capabilities. There seems to be too little opportunity for students to develop mathematical thinking abilities in the school curricula.

(g)   There are several levels of formality in each of various 'languages' in school learning. For example, there is the language of the home, of peers, of the classroom, and of writing. In posing problems or in devising achievement test items, great care must be taken to choose an appropriate level of language formality. The language tests, for example, might be a little more formal than that of the classroom and, therefore, it may have an impact on achievement data. Similar statements can be made regarding the structure of test items and problems. The interaction of national language structure with problem structure is one that must be handled very carefully and, perhaps, is not well understood at this time.

(h) An area that needs to be examined carefully is the role of ability grouping in elementary school and achievement. Closely related to this is the role of cooperative learning. Cultural variables enter into this. In Japan, for example, there is little ability grouping and students generally do not want to be different from other students. The result is that more concern for low achievers is shown; in contrast, ability grouping is more prevalent in the US, and achievement is significantly lower when compared to Japanese counterparts.

(i) In teaching problem solving, teachers can probably teach students the *procedure* of problem solving objectively. However, teaching the *strategy* of problem solving objectively may be more difficult. It may be more subjective and the strategy may be linked to the teachers and their individuality. We all know, for example, teachers who can teach *their* strategy of problem solving to students, but it is not easily transferred to another teacher. Teaching strategy and aspects of personality may be interrelated.

(j) The role of key words in problem solving is an important one in both Japan and the US. Teachers in both countries use key words, but probably less so in Japan, especially as problems get progressively more difficult as students move from grade to grade. However, the key word approach is frequently used improperly. The role of key words is to help the problem solver recall a situation; too frequently the approach is used in a way which connects key words to an operation which is incorrect. By using key words it is possible to imagine a situation, and then change its image to a familiar model situation which can be represented by, say, manipulatives or teaching aids, and only then decide which operation is appropriate by that model. The key word approach can be very useful if the following sequence is followed:

$$\text{key word} \longrightarrow \text{situation} \longrightarrow \text{model} \longrightarrow \text{operation}$$

If the key words are directly associated only to the end step, they will likely lead to incorrect results.

The use of manipulative teaching materials is common in both countries. However, we have to distinguish between manipulatives such as buttons, coins and beads, and structured manipulatives such as Cuisenaire Rods and Dienes' Blocks. The former are used more commonly in both countries, at the elementary level, and the latter less so in both countries.

(k) Pattern–finding behaviour in problem solving is considered an important problem–solving ability. Japanese elementary school textbooks commonly incorporate pattern–finding problems; however, this is frequently done by providing tables with blank spaces to be filled in by pupils. Also, the values of the independent variable are systematically arranged. Thus, pupils need only fill in the spaces and look for a pattern.   Perhaps it would be more valuable to have pupils identify variables and make the arrangement of values a prerequisite to searching for a pattern.   Arranging data in a systematic way does not seem to come naturally to pupils, but is very important in the search for relationships.   Perhaps it is at elementary school level where this should begin.

(l)   Both Japanese and American mathematics educators believe that schooling does not provide nearly enough emphasis on developing mathematical thinking abilities.   The result is that children become accustomed to not thinking about problem situations; instead, they give relatively automatic responses which are frequently meaningless. The solution is to develop curricula and teaching approaches that incorporate *thinking in mathematics* as a *basis for its design*.

(m)   In solving problems, students too frequently are not aware of assumptions they are making nor of the role of modelling in problem solving.   We may reasonably expect that students of the upper elementary grades can begin to understand these things. Whether the onset of this ability is due to natural intellectual growth, as students get older, is an interesting question.   It is also one on which cross–cultural research will focus, to determine at what level of intellectual growth this occurs or can be developed – with or without intervention.   It is not uncommon that even slower students will show a spark of interest and enthusiasm when challenged with problems in which model making plays an important role.   The area of model making is one that is little researched, and the development by students of a mathematical model as a tool in problem solving is one that needs to be further explored by researchers, both with and without the use of a calculator or computer.

3.   FORMATION OF THE JAPAN–US COLLABORATIVE STUDY GROUP

After the Hawaii Seminar, Japan–US collaborative study entered a new phase.   The Japan–US Collaborative Study Group is now formed and consists of the following researchers.   In Japan, Tatsuro Miwa (University of Tsukuba) as Co–Principal Investigator, Yoshishige Sugiyama (Tokyo Gakugei University), Toshio Sawada (National Institute for Educational Research), Nobuhiko Nohda (University of Tsukuba), Yoshihiko Hashimoto (Yokohama National University), Tadao Ishida (University of Hiroshima), Shigeru Shimada (Tokyo Science University), Shigeo Yoshikawa (Joetsu University of Education), and Toshiko Kaji (Tokyo Gakugei University). In the US, Jerry P. Becker (Southern Illinois University at Carbondale) as Co–Principal Investigator, James W. Wilson (University of Georgia), Edward A. Silver (University of Pittsburgh), Mary Grace Kantowski (University of Florida), and Kenneth J. Travers (University of Illinois/Urbana–Champaign).

4. **FOCI OF CROSS-CULTURAL RESEARCH STEMMING FROM THE SEMINAR**

(a) *Common survey of students' problem-solving behaviours*

A survey of students' problem-solving behaviours will be done in both countries with analysis of data from cross-cultural viewpoints on non-routine problems, pattern-finding problems, and modelling problems.   Subjects will be selected from grade levels in each country as follows: primary and intermediate grades of the elementary school, one grade of the lower secondary school, one grade of the upper secondary school, pre-service teacher training students (college or university).

(b) *Mathematical problem-solving practices in the classroom*

In Japanese schools, especially in the elementary schools, much emphasis is placed on cooperative learning among students.   On the other hand, in the US a great deal of emphasis is placed on individual performance throughout the school years.   However, many mathematics educators believe that small cooperative group learning may enhance problem-solving performance significantly.   We propose to investigate problem-solving practices at several levels and in several modes of cooperative learning.

(c) *Teachers' beliefs and behaviours*

Researchers of both countries think it is important to examine teachers' beliefs and behaviours in the classroom and how these behaviours are reflected in students' problem-solving performance.   For example, value judgements of teachers and their teaching methods will influence classroom practices.   Questionnaires will be administered to teachers, and some teachers will be videotaped in classroom situations – both student and teacher behaviours will be observed and studied.

(d) *Pre-service education of teachers*

Analyses of the rationale, philosophy, and practices of pre-service training programs in two Japanese and two US universities will be done, with a focus on problem solving, its role in teacher preparation and how the teaching of problem solving is fostered and promoted.

(e) *Casestudies based on the results of the Second International Mathematics Study (SIMS) of the IEA*

The results of SIMS reveal various differences between Japan and the US with respect to, for example, student and teacher attitudes toward mathematics and its teaching.   Another example is the use of hand-held calculators, which may be related to teachers beliefs about mathematics teaching.   Casestudies based on the SIMS results will be done in both countries.

(f) *Comparative study of mathematical problem solving in textbooks of Japan and the US*

A comparative study of the treatment of problem solving in leading textbooks of both countries will be carried out.

(g) *Implicit assumptions in algebra story problems*
The role of implicit assumptions and tacit knowledge in the solution
of problems will be studied.   This will involve study of students'
translation of ordinary language sentences, by which story problems
are presented, into sentences with algebraic expressions (equations).

5.   PRELIMINARY RESULTS OF RESEARCH

Collaborative research has recently started and we want to show one
example of a number of problems already used in preliminary research.
In 1987, N. Nohda and J. Becker studied US students' problem–solving
performance on the three types of problems described above.   Data was
collected in a similar fashion by N. Nohda in Japan.

Subjects in the research were students of third, fifth, eighth and
eleventh grades of public schools in Japan (Ibaraki Prefecture) and the
US (near Carbondale, Illinois).   Schools in both countries are situated in
similar rural localities.   Administration of the problems was carried out
by classroom teachers connected to the University of Tsukuba and
participating in a problem–solving project at Southern Illinois University.
Teachers were asked to do the following.

(a) In each class, the teacher distributed problem sheets to students and
read the problems to them, but did not give any explanation of the
problems nor any hints for their solutions.

(b) For each problem, students were given fifteen minutes to work.
The following problem is given here for illustration.

MARBLE PROBLEM

Three students, A, B and C throw five marbles that come to rest
as in the figures above.   In this game, the student with the smallest
scatter of marbles is the winner.   To determine the winner, we will
need to have some numerical way of measuring the scatter of the
marbles.

(a)   Think about this situation from various points of view and
write down different ways of indicating the degree of
scattering.

(b)   Which one do you think is the best?

The methods proposed by students to measure the scattering were classified and evaluated as follows (see Nohda 1987):
(a) visual observation (score 1),
(b) area of the polygon made by connecting the points (score 2),
(c) sum of the lengths of line segments connecting points (score 3),
(d) radius of the smallest circle containing all the points (score 4),
(e) other approaches.

Though the number of subjects was not large, we can see the following general results (see N. Nohda).
1. Japanese students gave a variety of approaches in the lower grades whereas a variety of ways for US students did not emerge until the upper grades. Japanese students proposed more ways of solving the problem in the lower grades and fewer in the upper grades than US counterparts. In contrast, US students proposed more ways in the upper grades. There were only small differences related to gender: in Japan, boys proposed somewhat more of a variety of approaches overall, and in the US intermediate grade girls showed somewhat more of a variety. We have no explanation yet of these preliminary results.
2. Generally speaking, lower–grade students in both countries proposed the method of visual observation fairly commonly, but this diminished in the upper grades. Approaches proposed by eleventh–grade students, for example, reflected a higher level of mathematical thinking.
3. When the same problem situation was given to thirty American teachers of grades 4–6 and 7–12 in July 1988, a variety of approaches were proposed, nearly all of which indicated higher mathematical thinking. The teachers were uniformly excited about such a problem and expressed the view that more such open–ended problems should be incorporated into the curriculum.

We expect to report findings of research subsequent to this preliminary work in the future.

REFERENCES

Becker, J. P. and Miwa, T. (1987). *Proceedings of the US–Japan Seminar on Mathematical Problem Solving*. Southern Illinois University, Carbondale, IL.

Nohda, N. (1987). *A Comparative Study of Mathematical Problem Solving between Japanese and American Students – From a Cultural and Development View*. Mimeograph.

CHAPTER 26

Application Readiness with Fractions

C. Ormell
University of East Anglia, Norwich, UK

ABSTRACT
The author began research on Application Readiness[1] in conjunction with Dr Ibrahim Abdel–Ghany (Minia, Egypt) in 1981. In the most recent work the idea has been extended to a specific topic, fractions. Preliminary results seem to show that children are commonly even less application ready with their knowledge of fractions than they are with their knowledge of basic (natural number) arithmetic.

1. INTRODUCTION

A child may be said to be application ready with a topic T in mathematics when he or she is able to see – naturally, spontaneously and unprompted – instances in which T can be used beneficially as a modelling device. To be application ready with T is to have the topic T as part of one's familiar problem–solving repertoire. If, as is widely agreed nowadays, we teach mathematics to children *chiefly* on account of its modelling and problem–solving capabilities, then it is clear that children need to be application ready with all their mathematical knowledge. To teach children mathematics of a problem–solving variety and in a problem–solving style *without* ensuring that they are actually in a state of application readiness with it is a major educational absurdity. If we introduce children to mathematics because it is an all–purpose, predictive and clarifying instrument in life and employment (as it is), then we must, of course, seek to introduce them to this subject in a form and at a level of skill at which it can actually work for them in this way. To leave children's knowledge of applicable maths in a form which is *in*applicable in practice, would be seriously to mislead and to disillusion children.   Having adopted the aim of making children's mathematics applicable to the real world, we ought, if we are not to be parties to a cruel tease, to ensure that it is *actually* a source of the pleasure which comes from being able to predict and clarify situations of life.

However, while it is not difficult to convince oneself of this argument in a general way, it is less easy to convert such an intention into a specific, practical procedure in relation to a particular mathematical topic, say fractions.   A host of difficult questions immediately present themselves, such as the following.

- Is there a single, canonical way to apply fractions to real situations?
- If there is, what would *examples* of this look like?
- What knowledge of fractions would be involved?
- If there is no single, canonical, way to apply fractions, what range of ways exist, and which of these are needed by the average child?
- What would examples of unprompted opportunities to use fractions in this *relevant* way look like?

Behind the last two questions there is the more radical question of whether, in this age of *calculators*, fractions are needed by anyone in ordinary life anyway.

Unless we can answer this question in the affirmative, there will be little point in proceeding to develop application readiness tests for fractions.   If fractions are not usefully applicable today in situations of ordinary life, the question whether a child is ready to apply them to situations of ordinary life will cease to have any importance.   So our first task must be to survey the extent, nature and importance of the applicability of fractions in modern life.

2.   THE APPLICABILITY OF FRACTIONS

With the arrival of the electronic calculator the role (ie applicability) of fractions in ordinary life has clearly diminished.   However, even though it has diminished, it has not dwindled to zero.   We tend only to use the simplest fractions like $1/2$, $1/4$, $1/3$, $1/5$, and to use them when either we are judging a quantity by eye, or where we know that a sharing process has been carried out.

We would say that a glass was 'a quarter full' rather than '0.25 full';   if we knew that someone's estate had been divided equally between five children on his death, we would say that 'they each received a fifth', not that 'they each received 0.2' of the estate.

Where an entity is already subdivided into parts (say, an hour) we would tend to use the units available rather than use fractions − apart from $1/4$, $1/2$ and $3/4$.   Thus 'a fifth of an hour' sounds much less natural than 'twelve minutes'.   However, we *might* say 'a fifth of a second' because there is no familiar subdivision of seconds.

This, then, is the area of experience to which fully numerate people tend to apply their knowledge of fractions today.   Fractions are, of course, in a sense much more widely applicable than this.   In the post–calculator era, though, many of these theoretical applications have ceased to be convenient or to have a point.   So we may call the area of applications sketched above the *current defacto* (CDF) area of applications.   Our main concern is therefore to inquire whether children are typically achieving a state of application readiness with fractions in the CDF area of applications.

The applications we are concerned with are of two types: (i) the simple use of fractional descriptions (as in spotting that $7\frac{1}{2}$ minutes is $1/8$th of an hour), (ii) compound applications where the operations +, −, ×, applied to fractional descriptions, lead us to a fractional conclusion, or a natural number conclusion.

3.   TESTING FOR APPLICATION READINESS

The method used has been similar to that which was applied to basic arithmetic (Ormel 1988), namely, that of giving children stories with mistakes, some of the mistakes being the failure of the childish characters in the story to resolve situations using simple fractional descriptions and operations.

Such latent storytests are written to a formula.   Each storytest contains two child characters, a boy and a girl, sometimes referred to as twins.   In the course of a short episode they make a number of verbal, informational and mathematical mistakes.   The children who are set these tests read the story, and then look for mistakes made by the childish characters.   The stories themselves contain no explicit *signals* about the need to introduce fractions in any way.   No specific fractions are mentioned in the tests, nor is the general term fraction used.   It is a condition for the administration of these tests that the children should not know or believe that they are being tested specifically for maths. We may call the lack of signalling of specific questions, *question anonymity*, and the lack of signalling of any mathematical–testing intent, *subject anonymity*.

During the Summer of 1988 four such latent storytests have been written by the author.

The first concerns a family with four cats which is sent three free tins of cat food.   An attempt is made by the children to let three of the cats have a tin each and to exclude the eldest cat: but the eldest cat joins in and eats a quarter of a tin before being stopped.   The crunch question is at the end, where the child characters have been offered another free tin of the catfood if they can fill in a questionnaire saying exactly how much each cat has eaten ($1/4$, $11/12$, $11/12$, $11/12$ of a tin). The childish characters duck this question, so their main mathematical mistake is that they failed to answer it.

The second story is about children who visit their mother in hospital.   She is offered tea on three occasions during the day, but only drinks $1/3$, $1/6$ and $1/2$ of a cup on each occasion.   Eventually the night nurse comes in and asks the child characters whether their mother has drunk a whole cup of tea altogether. They feel that they are being disloyal to their mother in revealing their impression that she hasn't drunk a whole cup altogether. This is the crunch mistake.

4. PRELIMINARY RESULTS

During the Summer of 1988 the author tried out these latent storytests on seven groups of children, totalling 173 children. Some children were tested twice, so that there were 222 test–response scripts.

The overall pattern of results of these trials is quite similar to that found in the case of the trials of the earlier latent storytests for detecting application readiness in basic arithmetic (1988). If we exclude two groups on a course for gifted children and one top set of third years in an above–average Middle School, there were only ten children out of 152 who spotted the central point at which an application of knowledge of fractions would make a dramatic impact on the story – a 6.6% success rate.

Virtually all these children could see that knowledge of fraction addition was applicable *after* it had been pointed out to them, but they did not spot it by themselves when they were positively engaged in the search for mistakes made by the characters in the story.

In one school *none* of the children in a group of 32 children spotted the main, compound application of fractional knowledge in the storytest. (Several spotted the minor, simple application of fractions involved in dividing three tins of catfood between four cats.) We have no reason to believe that these results are not broadly indicative of the general level of performance of children in this age band (11–13). In the case of the above–average Middle School, the children were first taken through a previous latent storytest and *shown* the way in which an opportunity to use fractional knowledge was buried in the circumstances of the episode. The resulting score of 38%, achieved after a mere 15 minutes of instruction on these lines, is suggestive of the hypothesis that children could be trained to reach a much higher level of alertness in these matters (ie application readiness) than they commonly achieve today – and by means of only a relatively small teaching effort specifically directed to this objective. The scores of the children on the gifted children course rose by 18% after two hours mixed work on storymaths of various kinds. In this case the final score (71.7%) may appear to be satisfactory, but this score disguises the fact that *seven* 'gifted' children either did *not* spot an application of fractional addition at all or, in one case, only partially spotted it.

Table of Results

Number	Successes	% Success	Notes
16	1	6.3	School A Yr 4
27	3	11.1	School A Yr 3 Visit 1
26	2	7.7	School A Yr 3 Visit 2
32	0	0	School B
21	8	38.1	School C
24	1	4	School D
27	3	12	School E (Secondary)
26	14	53.8	Gifted course day 1
23	16.5	71.7	Gifted course day 3
222	48.5	21.8%	

5.   INTERPRETATION OF THE PRELIMINARY RESULTS

As remarked above, there is a distinct difference between the results in School C (plus those for the Gifted course) and those of the remaining groups.   It is evident that the sample of children tested in schools A, B, D and E contains very few who have achieved application readiness with fractions in the CDF area.   This result is entirely consistent with the experience of many teachers, who have found (and hence come to expect) that only a small fraction of children acquire sufficient confidence in a topic such as fractions to become proactive in their use of it.

What is being suggested here is that this is not a defensible stance on the issue of application readiness.   Application readiness is, I suggest, *needed* by *all* children, both for intrinsic reasons (in terms of the children's gain in competence in handling practical matters) and in order to allow the children to justify the effort spent on learning mathematics to themselves.   This argument applies to the topics in the Cockcroft Report's *Foundation List* – those topics which every individual needs to survive in the modern world.   It applies with its greatest force, probably, to basic arithmetic, fractions, decimals, percentage and proportion, though in each case only a *part* of the syllabus area customarily marked out by these words may be needed by the child in a proactive form.

This argument is essentially *a priori* – relative to educational concerns.   It says, roughly, that individuals will lead happier lives if they acquire this mental accomplishment: it also presupposes that the accomplishment is *attainable* by the average child.

The issue of *attainability* is clearly a major one, and we certainly cannot beg the answer to this question. Much more research is needed to establish the extent of the attainability (of application readiness) in child populations.

Given the ideological nature of much of what we call education, however, the issue of attainability is closely tied to that of desirability. When teachers widely perceive the *desirability* of children attaining a state of application readiness in their Foundation topics, they are likely to *find* ways to encourage children to become application ready. It is like growing trees in the desert. This is an intrinsically difficult process, but once sufficient trees have been grown, the microclimate begins to change and at this point the growing of trees becomes distinctly easier.

Another major research area is that of the internal linguistic/dramatic structure of the storytests themselves. Much more needs to be done to specify stories with hidden mistakes which will serve as clear-cut detectors of application readiness. Two obvious criteria are that the stories should be fully readable and be interesting to the children. The author is satisfied that the tests used do meet these criteria, but the criteria themselves need to be more fully defined; the fact that they do meet these criteria needs to become explicitly verifiable. One of the by-products of the present research is the light it throws on the ability of children to respond to uncued, unprompted challenges.

Most of the children tested did write down a lot of real or imagined mistakes made by the childish characters in the stories. The involvement, indeed intense concentration, of the children on the stories is evident from the scripts. The teachers of the classes taken were, in all cases, fully convinced of the value of the activity – of the children searching intently for mistakes in the stories – independently of its research implications.

One problem which will have to be tackled is how to establish subject anonymity once a group of children become familiar with this form of test. A possible solution is that of writing bilateral tests which contain distractor mistakes from a specific subject, say geography, biology or history. Latent storytests of two kinds would be written; those testing for application readiness in subject A with mathematical distractors, and those testing for application readiness in mathematics with subject A distractors. Teachers would then enter *pacts* with their subject A colleague, so that each would set the tests in which their own subject supplied the *distractor* mistakes! This is probably the best we can do in the absence of a truly inter-disciplinary curriculum.

REFERENCES

Cockroft, W. (1982). Mathematics Counts. *HMSO*.

Ormell, C. P. (1972).   Mathematics, Science of Possibility.   *Int J Math Educ Sci Tech*, Vol **3**.

Ormell, C. P. (1988).   Application Readiness in Maths at 10/11.   *MAG*, Norwich.

[1]This is a reference to the author's work since 1964 to bring a heightened sense of applicability of mathematics to bear on the task of education. The idea of *projective* applications of mathematics, that is, the use of mathematics to model *possible* practical or theoretical developments, is explained in Ormell (1972).

CHAPTER 27

Recent Trends and Experiences in Applications and Modelling as Part of Upper Secondary Mathematics Instruction in Denmark

K. Hermann and B. Hirsberg
The Ministry of Education, Denmark

1.   THE DANISH SCHOOL SYSTEM

The Danish school system is divided into two subsystems, functioning as two separate systems as regards governing and teacher recruitment. Teaching of pupils between seven and 16 years of age (primary and lower secondary education) is provided by the *Folkeskole,* being a nine/ten year compulsory school, whereas non-vocational teaching of the 16–19 year old (upper secondary education) is provided mainly by the three year high school/grammar school, called the *Gymnasium*, admitting just under 30% of an age batch.   (Parallel to the Gymnasium, having the same governing body and using the same group of teachers, a two year education called *HF* (higher preparatory certificate) has been established, where certain subjects, particularly natural sciences, are taught at a level slightly below that of the Gymnasium.   Originally HF was intended to admit adults, who – when young – had never received any upper secondary education, but at present approximately 10% of an age batch is admitted.)

Instruction in the Gymnasium aims at further education and at providing general education.   The teachers hold university degrees, normally in two subjects, the levels of which correspond to the levels of a Master's degree.

2.   MATHEMATICS INSTRUCTION IN THE GYMNASIUM

2.1 Structure

The Gymnasium is composed of a linguistic and a mathematical stream.

As from August 1988, mathematics will no longer be compulsory for the students of the linguistic stream, but elements of the subject will form part of a science subject. Therefore, the following will deal exclusively with mathematics instruction in the mathematical stream of the Gymnasium.

Since 1960, and until this year, mathematics has been taught at two levels, called A and B, the number of lessons during the three year period being 5-5-6 and 5-3-3 respectively. During the first year teaching was common, but for the next two years the students were subdivided.

As from August 1988, the Gymnasium will be given a new structure involving, *inter alia,* that B-level in mathematics is achieved at the end of the first two years with five lessons a week, and A-level with a further five lessons during the third year.

The final examination consists of both an oral examination and a written examination prepared by the central authorities.

2.2 Content

In connection with the structural change the curricula of the individual subjects have been subject to a revision. As regards mathematics, the most marked change was to the effect that in addition to pure mathematical topics (at B-level: number theory, geometry, functions, differential calculus, statistics and probability; at A-level: integral calculus, differential equations, vector theory, geometry in two and three dimensions and computer mathematics) mathematics instruction should comprise the following three aspects.

1. The historical aspect, aimed at familiarising the students with elements of the history of mathematics and mathematics in cultural and social contexts.

2. The aspect of models and modelling, aimed at making the students familiar with building mathematical models as representations of reality. They should be given an idea of the potential and limitations in the application of mathematical models. In addition, the instruction given should enable them to carry out a not too complex modelling process.

3. The internal structure of mathematics, aimed at providing the students with an understanding of the modes of thought and methods characteristic of mathematics and their contribution to the development and structuring of mathematical topic areas.

3. APPLICATIONS AND MODELLING AS PART OF MATHEMATICS INSTRUCTION

The Direction for Teachers issued by the Ministry of Education, coming into force as from August 1988, describes the aspect of models and modelling as follows.

In not too complex contexts the students should be able to carry out a modelling process by themselves. By way of example the following contexts can be recommended.

*   Simple optimisation problems, simple geometry problem situations, where the geometric representation is not given in advance, and not too complex stochastic experiments.
*   Elements of model building and problems related to the building and application of mathematical models should be discussed. Recommended are the purpose of modelling, selection of the segments of reality under consideration, and idealisations thereof, mathematical representation including possible subsequent simplification and loss of information, verification problems.
*   In connection with the treatment of the main topics, examples of models should be introduced, for example linear and exponential growth models. In addition, for example, difference and/or differential equation models, linear programming or stochastic models can be included.
*   General, dynamic or stochastic simulation programs can be studied.
*   The above examples can also constitute self-contained instructional sequences. Likewise a major authentic model (parts thereof) can be studied, for example a physical, economic or ecological model. The social impacts of such models should be discussed.

The change in the curriculum is the result of a long process comprising curriculum innovation and experimental instruction, inspiration from mathematics educators in universities, discussions at the individual schools and with the Association of Mathematics Teachers, teachers' in-service education and so on.

In the following, the main points in this process, the current state of mathematical modelling and applications will be outlined.

3.1 Background to the curriculum modification

From 1960 until the mid-1970s instruction was strongly influenced by the new mathematics movement, which had considerable impact on the mathematics curriculum introduced in 1960. Although the curriculum had been modified in 1971 – among other things the syllabus was cut down because no instruction was given on Saturdays any more – the content was still rigorously described through a number of purely mathematical topics. Likewise the written examination papers prepared by the central authorities did not include applications except for a few examples in calculus of probability.

In 1960 only 10% of an age batch went to the Gymnasium, and almost all of them continued into higher education. However, in 1975 admission had almost doubled, the students' backgrounds were more heterogeneous than before, and they continued into a wider spectrum of education and jobs than earlier. For several of them the Gymnasium became their final educational experience with mathematics.

Many teachers and students felt the upper secondary mathematics did not transgress its own borderlines and left a good deal to be desired.

Means to remedy the situation were discussed among ordinary teachers, association leaders and curriculum planners. Great interest was shown in introducing new aspects of mathematics, and in this connection particularly applications and modelling were the focal points.

The fact that the teachers of other subjects taught in the Gymnasium had simular difficulties and showed great interest in holistic instruction with interdisciplinary cooperation exerted influence on the situation too.

3.2 The innovation process

In the years that followed (1971–1988) a number of parallel activities were carried out, initiated by the Ministry of Education, the Association of Mathematics Teachers and the individual schools and groups of teachers.

3.2.1 Teachers' in-service education

Almost every year, at the beginning of this period, in-service courses (prepared by the Ministry of Education and the Association of Mathematics Teachers) were held. They dealt with mathematical applications, particularly within such fields as biology, economics, physics, sociology, and many teachers introduced such major application aspects in the 40-hour free topic at A-level, being the only place where the curriculum allowed for such instructional sequences.

Later courses emphasised the different aspects of mathematical modelling, *inter alia* courses where the participants were placed in a modelling situation.

During this period great interest was also shown in other types of courses; courses on mathematics education discussing the further development of mathematics instruction.

Also, the written examination papers, which had a substantial impact on mathematics instruction, were discussed. At one particular course new types of problems, quite open problems, were presented. These were problems which were intended to make the students carry out one or more of the following processes themselves: formulate the problem, set up a mathematical model and solve the mathematical problem, evaluate the mathematical model applied, evaluate the premises and results. Today, the written examination papers do not comprise open problems, but the debate contributed to changing the papers so as to include more real applicational problems, and the open problems were introduced in the mathematics lessons.

An example of such an open problem is given in the following.

Using your knowledge of mathematics, you are requested to evaluate the statement made in the newspaper extract below about returning verdicts in the way described.

"Jury verdicts do not require unanimity, but a verdict of guilty can be returned with a majority vote of at least 8–4. The vote of a single, inexperienced juror can thus be decisive for the verdict returned, ie whether the verdict will be of not guilty or it will be a sentence of long term imprisonment. It should be stressed that if the jurors made their decision by tossing a coin – and nobody would call that justice nowadays – the verdict would become a verdict of guilty with eight votes to four in 12% or $1/8$ of the cases. The verdict returned may thus be a matter of chance. For the verdict not to be a matter of chance the number of votes has to be 10 or more of the 12 votes cast, ie there has to be a significance as it is called in statistics. "

3.2.2   Curriculum innovation and experimental instruction

At the end of the 1970s, comprehensive, innovative teaching programmes were launched at a number of schools aiming, among other things, at including major, authentic, mathematical applications and/or mathematical modelling in mathematics instruction.

Some of these innovative teaching programmes took place as interdisciplinary cooperation or were combined with out–of–school activities such as excursions to various factories, agricultural research stations and the like in order to study their applications of mathematics.

Such sequences were only made possible by exemptions being granted from the existing curriculum and formal regulations, and individual schools and teachers were granted such exemptions. In addition, exemptions were granted to teachers who wanted to introduce new topic areas.

The many individual exemptions, and the different written examination papers to match, made the situation most unfortunate, and, in 1981, the Ministry of Education restricted the number of individual exemptions and granted a standard dispensation. This dispensation gave permission to leave out some of the purely mathematical topics to be replaced with, for example, mathematical applications or mathematical modelling, elements of the history of mathematics, or mathematics in historical, cultural, philosophical or social contexts.

When the standard dispensation had run for a couple of years, a standard experimental mathematics curriculum was published in 1983–84 by the Ministry of Education, following discussions held at a number of meetings of teachers all over the country and taking into account the resulting proposals for revision. It was made possible for interested teachers to adopt the latter curriculum. The content of the standard experimental mathematics curriculum was very similar to that of the curriculum coming into force in August 1988.

After the publication of the standard experimental mathematics curriculum, activities in textbook writing increased, and new textbook systems, adapted to the curriculum, were published. In addition, textbooks particularly aiming at instruction in the aspects of models and modelling were published. Among others, the following titles can be mentioned: Model Talk, The Labour Market Model in SMEC III (the simulation model of the economic counsel), Mathematical Fishing Models, Mathematical Models in Economics, Computer Models, System Dynamics, Spontaneous Phenomena, and Linear Programming. In this connection, various computer programs were developed of which some illustrated a specific mathematical model and some were general tools, for example programs for the treatment of differential equation systems.

The aspect of models and modelling has been treated both in connection with compulsory, purely mathematical topics and in specially planned sequences, where new mathematical topics or the students' existing knowledge of mathematics could be incorporated.

In 1987 more than 50% of the students were given instruction according to the experimental mathematics curriculum, and reports from teachers, who adopted it in 1986/87, show that the aspect of models and modelling was included in connection with the following mathematical topics: exponential functions, differential equations, trigonometry, probability calculus, linear programming, recursion and graph theory. Furthermore, it was noted that applications focused on economics, ecology and genetics.

3.3 Examples illustrating the change which took place from 1971–88

An example illustrative of the development is the way textbooks treat exponential functions.

In the 1970s, the point of departure for the treatment of exponential functions was that they are inverse functions to logarithmic functions, and the emphasis was on the theoretical aspect. All students at A–level knew, for instance, that a^x is an isomorphism $(R,+) \to (R_+,.)$, but few had been taught that exponential functions had something to do with constant growth in percentages.

In textbooks published around 1987, the point of departure for instruction in logarithmic and exponential functions is exponential functions, and many efforts are devoted to explaining exponential functions as mathematical models of specific growth forms.

Textbooks published around 1970 contained very few applications of the mathematical topics, whereas just one of the new systems from 1987 describes applications such as loan raising, cost minimisation, tides, parachute jumping, conversion of alcohol in the human body, dosage of medicine in connection with injections and epidemics. Also, the usual areas of applications were included: models within probability theory, applications in physics and astronomy and linear, exponential and logistic models of growth.

Finally, some examples from the written examination papers will be given.

In the beginning of the 1970s the students' knowledge of exponential functions was tested in the usual, purely mathematical, way. Recent years' written examination papers also contain purely mathematical problems (approximately 50%), but also problems like the following are included.

A–level 1984:
Lead contamination of grass is mainly due to combustion of petrol in cars. The lead content in grass near a road has turned out, on a rough estimate, to decrease exponentially with the distance from the roadside, and the half–value distance is 15m. At a distance of 8m from the roadside of a motorway, a lead content of 50mg per kg of grass was measured.
*   Determine the lead content of the grass at the roadside.
*   How far away from the roadside should cows graze when an EC requirement for a maximum lead content of 10mg per kg of forage grass has to be fulfilled?
The lead content of the grass (measured in mg per kg) is denoted f(x) at a distance x (measured in metres) from the roadside. The average lead content of the 50 metres closest to the road is consequently determined by

$$\frac{1}{50} \int_0^{50} f(x)\,dx.$$

Calculate the average content.

B–level 1988:
The following excerpt is from the book entitled Principles of Sterilization and Disinfection by Knud Skadhauge, and it shows the radiation resistance of various bacteria.

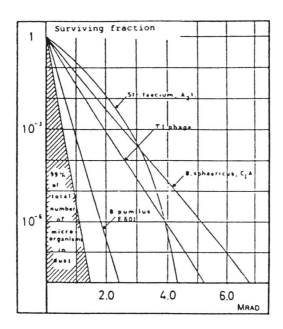

Figure 15 : Strålingsresistensen af forskellige bakterier. (fra: Ebba A. Christensen, USP, Open Conference on Biological Indicators for Society Assurance, Arlington, Virginia, 1970)

• Give arguments substantiating that the fraction of surviving bacteria of the species B. sphaericus is an exponentially decreasing function of the radiation dose, measured in M rads, and find an expression for this function.
• Calculate the half-value doses of B. sphaericus.
• Calculate the radiation dose necessary to reduce the fraction of surviving B. sphaericus to 1/500.

As will appear from the above examples, the students are not required to mathematise problems from the written examination paper themselves. Most teachers find that such requirements are incompatible with the time limit of four hours set for a written examination, where several sides of the students' mathematical knowledge are to be tested.

The new curriculum takes effect from August 1988 and, thereby, mathematical modelling is made compulsory. Developments have been characterised by an open dialogue between the authorities and groups of teachers, and debates and meetings held to inform teachers about the new curriculum show that they agree that the introduction of the new curriculum has given applications and modelling a fair position in mathematics instruction.

CHAPTER 28

Survey of the Present State, Recent Developments and Important Trends of Modelling and Applications in FR Germany

G. Kaiser—Messmer
University of Kassel, FR Germany

ABSTRACT

This survey starts with a description of theoretical developments concerning applications and modelling, discriminating three trends in the German—speaking discussion. Curriculum projects are then described and the situation concerning teaching materials is analysed, followed by an overview of the results of empirical investigations on applications and modelling. The survey closes with an analysis of the role and place of applications and modelling examples in the syllabuses, and the situation in schools.

1. THEORETICAL DEVELOPMENTS

In a simplified view, it is possible to distinguish *three trends* in the *German* educational debate on applications and modelling in mathematics teaching, including the *Austrian* discussion as well:

1. the *emancipatory* trend (exponents, among others, are D. Volk, P. Damerow and C. Keitel),
2. the *science—oriented* trend (exponents, among others, are H. G. Steiner and A. Engel),
3. the *integrating* trend (exponents, among others, are W. Blum, R. Fischer & G. Malle, H. Schupp, H. Winter and E. Wittmann).

These trends differ from each other primarily by the *goals* connected with applications and modelling in mathematics education. Thus the emancipatory trend starts from an interest in *emancipation* as the main orientation, and chiefly claims *utilitarian* goals. Mathematics teaching should enable the students to act in a rational and autonomous manner in actual or future real life situations. Yet this trend differs by its explicit theoretical basis from pragmatic approaches, largely accepted in the English—language area, which restrict themselves to utilitarian goals. The science—oriented trend places the *science* of mathematics and its *epistemology* in the foreground and requires mainly *methodological* and epistemological experiences by the students. This trend emphasises the necessity to gain insight into the modelling and mathematisation processes whereby some, especially Engel, claim goals referring to mathematics as a subject matter. The integrating trend demands a *balanced* relation between utilitarian, methodological/epistemological and mathematical goals. This trend is strongly influenced by the general pedagogical goals for mathematics teaching, formulated by Winter (1975).

The German discussion was *polarised* from its beginning until the mid—eighties, an important part of the theoretical background referring to the emancipatory and especially to the science—oriented trends. In the last few years ideas belonging to the *integrating* trend have become *more relevant*, in the course of which some of the extreme positions have lost their importance and others changed their conceptions, mainly by including utilitarian goals (eg the Austrians Fischer & Malle).

In summary, it is now a consensus, largely accepted in the educational debate, that applications and modelling should achieve the following goals:

- *utilitarian* or *pragmatic* goals – promotion of abilities to master everyday life and situations from other subjects or sciences;
- *methodological* goals – teaching of abilities to translate between the real world and mathematics, especially modelling abilities and abilities to apply mathematics in real world situations;
- goals referring to *mathematics* – increase of the motivation of the students to do mathematics, promotion of the long term retention and enhancement of the comprehension of mathematics;
- general *pedagogical* goals like the promotion of creativity or problem—solving abilities;
- *scientific theory—oriented* goals that aim to show a balanced picture of mathematics as a cultural and social phenomenon.

The three strands of the discussion also develop different thinking on the *relevance* of *applications* and *modelling in mathematics teaching*, and the *ways* of considering them. The emancipatory trend abandons the mathematical structure as the orienting guide of mathematics teaching. Volk, especially, concludes from the goals of mathematics education that the treatment of mathematical methods should serve to develop the ability to act rationally in real—world situations, and therefore has to happen in the context of real—world problems with mathematics as an aid and a tool. From this basis he states the demand for an *interdisciplinary* or cross–disciplinary teaching. For the science–oriented trend (especially as propounded by Steiner) applications and modelling should form the basis of the modelling and mathematisation processes, leading to *mathematised mathematical theory* too.

The integrating trend formulates an intermediate position between these two approaches, such as giving applications and modelling a high priority in mathematics teaching, without accepting them to be the only criteria for the selection of mathematical content and the structure of mathematics teaching. Winter, for example, requires for the secondary level a *cooperation* between *mathematics* and *other teaching subjects* in projects or *project–oriented* coursework, which is also demanded by Fischer & Malle. Alongside these global applications and modelling examples, Blum· demands an enrichment of the mathematical content, made compulsory by the syllabuses through smaller applications and references to real–world situations.

The thinking on the *usage* of *mathematics* and the mathematical *activities* of the *students* has changed considerably in the few last years. It is now largely consensus, indeed, that modelling and mathematisation processes have to be considered in mathematics teaching, but an extensive discussion comparable to those of the English–language area has not taken place. So far, it is largely accepted that the modelling process and the teaching of modelling abilities have to be stressed, as against the teaching of standard models. In contrast to the English–speaking debate, Fischer & Malle as well as Blum, emphasise the necessity to explicitly *reflect* upon the *modelling process* and the underlying *epistemological questions* in mathematics teaching.

Altogether the German–speaking discussion considers and refers to the international debate; ideas on modelling and applications have been developed continuing some typical German traditions, especially the consideration of practical applied arithmetic (so called *Sachrechnen*) and the emancipatory pedagogy (for details, see Kaiser–Messmer 1986).

Discussions and reflections on the possibilities afforded by the use of *computers* also take place, leading to the hope that computers will facilitate the treatment of more realistic and complex applications and modelling processes in mathematics teaching. Nevertheless, until now, these considerations have not had an impact on the theoretical debate on application–oriented mathematics teaching.

2. CURRICULUM PROJECTS AND MATERIALS FOR APPLICATIONS AND MODELLING IN MATHEMATICS TEACHING

Many *teaching materials* for all ages and levels have been developed in the last few years by several curriculum projects, which have been founded since the mid–seventies.

One well known project is the *Mathematik–Unterrichts–Einheiten–*

Datei (file on mathematics teaching units), organised and supported by teachers themselves, which is strongly connected with the emancipatory trend of the educational discussion and has developed numerous teaching materials related to sociocritical and environmental problems (for details see Böer & Meyer–Lerch in Blum et al 1989).   More official projects are the experimental schools *Laborschule* and *Oberstufenkolleg Bielefeld*, which emphasise interdisciplinary and project–oriented approaches for mathematics teaching (for details see Schluckebier & Stanik, Effe–Stumpf & Kemper, Gerull, Schülert & Vohmann, in Blum et al 1989). Furthermore, several states of the Federal Republic run special teacher–training courses on project–oriented coursework in the lower and upper secondary level, which have developed many teaching materials.   Real–world materials for weaker and lower ability pupils have been developed at the University of Bremen, especially for stochastics teaching at the University of Saarbrücken (for details see Jäger & Schupp 1983).

At the *University* of *Kassel*, German and international literature on application–oriented mathematics teaching has been collected and analysed since the mid seventies.   The aims of this project are, on the one hand to provide an overview of the literature relevant for practising teachers and pre–service and in–service teacher training and, on the other hand, to conduct a survey on the developmental trends concerning theoretical analysis and general considerations related to applications and modelling in mathematics teaching (see Kaiser, Blum & Schober 1982).   This analysis shows, amongst other things, a change in the type of the teaching materials developed in the last few years.   In the seventies, mainly project–oriented teaching materials were developed, with extensive real–world problems as a starting point.   In contrast, a large number of materials have been produced in the last years which propose real–world examples within the syllabuses for mathematics, serving the introduction of new concepts or the exercise of already introduced methods. Altogether, most of these materials are characterised by their concentration on the content.

3.   EMPIRICAL RESEARCH

So far, only a few *empirical investigations* have been carried out on applications and modelling, corresponding to a deficit concerning empirical research in the (West) German discussion on mathematics education, with a concentration and orientation toward reflections on the content.   Most of the empirical investigations on applications and modelling (eg Bender, Niegemann, or Strässer in Blum et al 1989) are concerned with the problem of how to teach skills to apply mathematics in real–world situations.   These investigations point out that it is necessary to learn how to apply mathematics in real–world situations, especially with unexpected mathematical methods or varying extra–mathematical context, because most of the students are not able to apply intra–mathematically learned concepts in real–world situations.   The major problems of the students while using mathematics to solve real-world problems are missing real-world knowledge and their haste to use well-known mathematical methods.   This is partly caused by teachers, who themselves do not take the world outside mathematics seriously, and disregard the interpretation phase of the mathematical solution in the real context.

In my own empirical research, carried out in recent years, I have examined more globally the possibilities of realising the goals of an application-oriented mathematics teacher (see Kaiser-Messmer in Blum et al 1989).   Six *casestudies* on calculus courses in the upper secondary level and additional investigations have yielded (among others) the following *results*.

- The ability to master everyday life situations using mathematical methods can be reached extensively by most students, even the weaker ones.   However, this ability has to be taught and cannot simply be expected to transfer from intra-mathematical teaching.

- Modelling abilities are only attainable in long-term learning processes, mainly with highly motivated and talented students. Especially the ability to interpret and apply mathematical concepts in unknown real-world situations is strongly promoted by an application-oriented introduction.   Barriers to the promotion of modelling skills are schematic-thinking procedures and adherence to recipes.

- Application-oriented mathematics teaching increases the motivation and interest in mathematics of nearly all the students.   The attitudes of some of the students with negative attitudes towards mathematics can be improved, but only as a long-term effect.

4.   SYLLABUSES AND SITUATION IN SCHOOLS

The development of the syllabus is the responsibility of each of the 11 federal states of the FRG.   Thus, there are different syllabuses for each state.   I here restrict myself to the lower and the upper secondary level.

4.1   Syllabuses for the lower secondary level (age 11-16)

There are different syllabuses for the three types of school into which the lower secondary level is divided (Hauptschule, Realschule, Gymnasium).

The syllabuses of the *Hauptschule* (for the so-called lowest ability pupils) are largely application-oriented, as they were before the new math movement.   The focus of applications is on practical applied arithmetic, comprising newer mathematical fields like descriptive statistics as well as the classical fields like percentages, interest and so on.   Some syllabuses link up with reform/pedagogical ideas and prescribe the treatment of extra-mathematically structured fields for the graduation classes, which serve a more extra-mathematical purpose.   The topics of these fields are usually taken from business and commerce.   Some of the syllabuses recently developed require that the relations between real world and model be reflected, and the students be enabled to mathematise.

On principle, the syllabuses of the *Realschule* (for the so-called middle ability pupils) are less application-oriented than those of the Hauptschule, but with some differences between the different states. As in the case of the Hauptschule, applications tend to gain in importance by taking recourse to parts of traditional practical applied arithmetic and by including modern mathematical fields. Extra-mathematically structured teaching units are prescribed but only in a few syllabuses, and often in the shape of additions, especially optional subject matter.

The syllabuses of the *Gymnasium* (for the so-called highest ability pupils) are even less application-oriented than those of the Realschule. Practical applied arithmetic remains very often confined to its formal/mathematical aspects. Some syllabuses developed for the *comprehensive schools*, of which there are only a few existing, tend to counteract this; thus, some syllabuses strive to assign a higher status to practical applied arithmetic and to introduce extra-mathematically structured teaching units towards the end of the lower secondary level.

4.2 Syllabuses for the upper secondary level (age 16–19)

The dominant feature of this school level with regard to the mathematical subject to be taught are the three pillars – calculus, linear algebra/analytical geometry, and stochastics – but there are remarkable differences between the syllabuses existing for each federal state. One extreme is represented by syllabuses essentially *limited* to *subject matter* catalogues and detailed learning goals, with only a very few referring to applications and modelling examples. Other syllabuses emphasise the necessity to *integrate applications* and modelling examples into the mathematics courses and provide encouragement expressly to carry out projects in the classroom, if there is sufficient time. Some of these syllabuses require, in their general educational goals and principles, the school to teach the ability to develop mathematical models in simple cases and to gain insight in some important applicational fields of mathematics.

4.3 The situation in schools

Concerning the situation of applications and modelling examples in schools, there is a considerable *gap* between the far-reaching acceptance of applications and modelling in the educational discussion and, even among teachers, a low actual importance in schools. A representative *survey* for the upper secondary level, carried out in 1982, pointed out that on average only approximately 15% of the whole teaching time is devoted to applications (in the broadest sense).

- About 40% of the teachers rarely start out from real-world problems when introducing a new topic, about 40% do this sometimes.

- About 30% of the teachers seldom use word problems to exercise mathematical methods, only about 15% do this regularly.
- About 90% of the teachers rarely present problems which require modelling processes in order to solve the problem, only about 5% do this often.

Concerning the other school levels, the situation is substantially better, according to the different role and relevance of applications prescribed by the syllabuses of the different school types (for details see Beck, Biehler & Kaiser 1983).

This gap is not really caused by teaching materials. The reasons lie chiefly in the crowded syllabus, and the higher demands of applications and modelling examples for teacher and students. Furthermore the lack of employment of young teachers since the beginning of the eighties has, meanwhile, lead to an aged teaching staff and the slowing down of the pedagogical innovation processes. However, it is hoped that the growing awareness of the necessity to transfer the high- level educational discussion on applications and modelling in pre-service and in-service teacher training will have an important impact on schools in the next few years.

REFERENCES

Beck, U., Biehler, R. and Kaiser, G. (1983). National report: Federal Republic of Germany. In Burkhardt, H. (ed) *An international review of applications in school mathematics – the elusive El Dorado*. Columbus, Ohio: ERIC, 68–93.

Bender, P. (1988). Eine neue Untersuchung zur sachmathematischen Kompetenz von Viert- und Fünftklässlern. In Bender, P. (ed) *Mathematikdidaktik: Theorie und Praxis – Festschrift für Heinrich Winter*. Berlin: Cornelsen, 15–28.

Blum, W. (1985). Anwendungsorientierter Mathematikunterricht in der didaktischen Diskussion. *Mathematische Semesterberichte*, **32**, 195–232.

Blum, W. et al (ed) (1989). *Applications and Modelling in Learning and Teaching Mathematics*. Chichester: Ellis Horwood.

Damerow, P. (1984). Mathematics for all – ideas, problems, implications. *Zentralblatt für Didaktik der Mathematik*, **16**, 81–85.

Engel, A. (1982). Statistik auf der Schule: Ideen und Beispiele aus neuerer Zeit. *Der Mathematikunterricht*, **28**, 57–85.

Fischer, R. and Malle, G. (under cooperation of Bürger, H.) (1985). *Mensch und Mathematik*. Mannheim: Bibliographisches Institut.

Jäger, J. and Schupp, H. (1983). *Curriculum Stochastik in der Hauptschule*. Paderborn: Schöningh.

Kaiser, G., Blum, W. and Schober, M. (1982). *Dokumentation ausgewählter Literatur zum anwendungsorientierten Mathematikunterricht*. Karlsruhe: Fachinformationszentrum Energie, Physik, Mathematik. Addendum to appear 1990.

Kaiser—Messmer, G. (1986). *Anwendungen im Mathematikunterricht*. Vol 1, 2. Bad Salzdetfurth: Franzbecker.

Keitel, C. (1985). Mathematik für alle — ein Ziel; was sind die Ziele einer 'Mathematik für alle'? Hans Freudenthal zum 80. Geburtstag. *Zentralblatt für Didaktik der Mathematik*, **16**, 177–186.

Niegemann, H. (1983). Problemverstehen und Transfer: Theorieübersicht und ein empirischer Beitrag zur Frage der internen Bedingungen für die Anwendung mathematischen Fachwissens auf Textaufgaben. In Kötter, L. and Mandl, H. (ed) *Kognitive Prozesse und Unterricht*. Düsseldorf: Schwann, 233–261.

Schupp, H. (1989). Applied Mathematics Instruction in the Lower Secondary Level — Between Traditional and New Approaches. In Blum, W. et al (ed) *Applications and Modelling in Learning and Teaching Mathematics*. Chichester: Ellis Horwood.

Steiner, H. G. (1976). Zur Methodik des mathematisierenden Unterrichts. In Dörfler, W. and Fischer, R. (ed) *Anwendungsorientierte Mathematik in der Sekundarstufe II*. Klagenfurt: Heyn, 211–245.

Volk, D. (1979; 1980). Vol A: *Handlungsorientierende Unterrichtslehre am Beispiel Mathematikunterricht*. Vol B: *Zur Wissenschaftstheorie der Mathematik*. Bensheim: Päd—extra—Buchverlag.

Winter, H. (1975). Allgemeine Lernziele für den Mathematikunterricht? *Zentralblatt für Didaktik der Mathematik*, **7**, 106–116.

Wittmann, E. (1981). *Grundfragen des Mathematikunterrichts*. Braunschweig: Vieweg.

CHAPTER 29

Problem Solving, Modelling and Applications in the Finnish School Mathematics Curriculum 1970–1985

P. Kupari
Institute for Educational Research, University of Jyväskylä, Finland

ABSTRACT

In this article I shall discuss the development of the mathematics curriculum in the Finnish comprehensive school (grades 1–9, ages 7–16 years), which, in 1970, replaced the binary school system (elementary/secondary school). The presentation concentrates mainly on problem solving and applications, because modelling has not been discussed in the Finnish curricula as an independent activity. It is, however, mentioned in the curricula, and most often in connection with application and problem solving.

1. ON THE DEVELOPMENT OF THE COMPREHENSIVE SCHOOL MATHEMATICS CURRICULUM IN FINLAND 1970–1985

1.1 The 1970 curriculum of the comprehensive school

The report of the Committee on the Comprehensive School Curriculum introduced simultaneously both new mathematics and the new school system in 1970. The new curriculum emphasised the development of thinking skills in addition to cognitive skills but it did not, however, include details of subject curricula. At the same time the division of school mathematics into three – arithmetics, algebra and geometry – was removed and many teaching contents of algebra and geometry were set to lower grades.

The curriculum stated the following on application and

problem—solving (Peruskoulun opetussuunnitelmakomitean mietintö II 1970).

> "Pupils are instructed to acquire ways and means as well as methods needed in obtaining and applying mathematical knowledge independently ... For later studies and use of mathematics skills it is important that the pupil learns to work independently: he must get used to obtaining information in books, tables and other sources of mathematical knowledge, and learn to apply the information in practical situations as well as to control his own achievement".

One aim of the curricular reform was to change the overall structure of mathematics instruction so that the pupils' mathematical thinking would, through instruction based on understanding, develop continuously from the first grade on. However, the over—emphasised position of set theory and bad timing of the implementation of the reform only scratched the surface of the intended improvement. New mathematics aroused, at that time, more resentment than admiration, and according to some experts in the field the reform was leading mathematics instruction in an undesirable direction.

The new school form was probably one of the main causes for insecurity because the teachers had to acquaint themselves simultaneously with the new school system and the teaching of new mathematics subject matter. In the comprehensive school, the entire age cohort was taught together for two years more than previously. From the 7th grade onwards the pupils were given an opportunity to study in set groups (the general, middle and extensive sets). The effect was also made worse by an administrative change which broke the old teaching tradition. Class teachers were required to retain their pupils for two additional years and teach subject matter for which they had not been trained. Teachers of the upper level had to take responsibility for the first time over the entire age group, and could have the same pupils for only three years.

Some experts in this field thought optimistically that new mathematics would expand pupils' skills in problem solving. From the point of view of pupils' quantitative thinking, problem solving was seen as a key factor because it reflected a child's ability to understand the situation of a real problem, take advantage of something learned, and apply it. Especially, technical higher education expressed the importance of modelling in school mathematics.

In the writings of that period, different textbooks and application tasks were perhaps criticised most. The artificiality of tasks and the over—emphasis on formulae in the textbooks were criticised. According to an accepted expert, 'dubious characteristics' had been included in our textbooks. He characterised it as follows. "Very often application tasks include formulae where the numbers are marked mechanically and which have the mathematical sign indicating the operation beside them. When a pupil is faced by a similar task without the formula he cannot calculate, because the corresponding exercise is missing".

The appraisals of future curricular improvement considered the development of application and problem–solving skills as important aims. Also, the national tests of mathematics achievement by the Institute for Educational Research in the early 1970s revealed a need to intensify application instruction.   The attainment of the aims was considered possible by paying more attention (also in teaching practices) to formal objectives, not only to the presentation of more extensive subject matter.

1.2 Core objectives and basic subject matter

On the basis of feedback gained from the implementation of the 1970 curriculum, it became evident that the mathematics curriculum was too extensive in relation to the number of lessons available.   It was realised that the total number of mathematics lessons was smaller than in both of the previous school forms, which were replaced by the comprehensive school.   According to one study, the transfer to the comprehensive school system meant that the number of lessons decreased by about one fourth, that is by one entire school year of study.   Furthermore, teachers' experiences revealed that the (at that time fashionable) spiral principle was ill–suited to mathematics instruction.   The subject matter was presented in bits, there was little time and this resulted in superficial learning.

In 1975 the National Board of General Education (NBGE) set up a group for developing comprehensive school mathematics instruction.   A Proposal for Core Objectives and Basic Subject Matter was born as a result of the Group's work, and it was published in 1976.   Even as a proposal it clarified the interpretation of the curriculum which could be seen in textbooks and in the decrease of teachers' insecurity.   The proposal also presented some other facts to be considered, one of them being problem solving (Kouluhallitus 1976):

"The pupil should be instructed to observe problems in his environment which could be solved by mathematics ... A desire to solve problems should be aroused in the pupil.   The problems should be sufficiently interesting, important from the point of view of the pupil and not too difficult in the beginning."

In   different   quarters   (teachers,   textbook   publishers,   school administration), there existed a fairly unanimous opinion about the necessity of the core objectives and basic subject matter.   It was agreed that in this way a more solid basis for pupils' performance could be obtained and that low achievement could be simultaneously decreased.

What did the change then mean in application and problem solving? Concerning the textbooks and use of teaching time, there were two kinds of development.   On the one hand, after the set–theory hysteria of the early 1970s, teachers returned to the old ways;   they focused their attention on applications and increasing the number of tasks requiring applications.   As an example of this trend, the 8th grade textbooks, between 1974 and 1979, included over one third more application tasks (verbal problems) than before.   On the other hand, the decrease in the teaching time of mathematics and the stripping of the mathematics curriculum represented another kind of development. The result was, no doubt, that less time-consuming subject matter was taught, or attempts were made to accelerate the teaching process – the reduction of subject matter being aimed, in particular, at tasks related to concrete contexts and at the number of applications. The situation raised a dilemma: even though there was a relatively large number of application tasks in the textbooks, they could not be used in instruction (or there was no will to use them).

The position of applications in the curricula of the 1970s was also analysed in the study of Leino and Norlamo (1980). The researchers felt that, among others, the following factors were responsible for the underestimation of applications in teaching:
1. teachers' poor knowledge of practical applications,
2. pupils' unfamiliarity with applications and the fact that these applications are so detached from real situations,
3. the excessive time application activities require,
4. the difficulty of applications to many pupils.

In the study, vocational school teachers felt that the comprehensive school should increase instruction in applied mathematics (eg use of quantities and equations, ratio and proportionality, percentages). Vocational school instruction, in turn, should stress contents that are closely connected with real situations and emphasise readiness to apply mathematics in practice.

On the basis of comparisons made between different textbooks, vocational school books had plenty of applications, which could be taught in the comprehensive school. In the vocational school textbooks the proportion of purely mathematical tasks was about 30–60%, whereas in the comprehensive school textbooks the proportion of these tasks was as high as 70–90%.

The authors emphasised problem-oriented instruction, where time should be devoted to the proper study of mathematical situations in which pupils are interested, or otherwise value. Time should also be allocated to discussions of pupils' own suggestions for solutions. Further on the authors attempted to promote an increase in applications by providing an extensive and rich collection of application tasks in their study. The material was classified according to content area and aimed for use both in the comprehensive and upper secondary schools.

1.3 The 1982 comprehensive school mathematics syllabus
At the turn of the 1980s, problem solving became the slogan for the next reform. In the new 1982 comprehensive school mathematics syllabus, the goals were set to enhance new ways of instruction that would emphasise creativity and the development of thinking skills. Especially, these new endeavours were introduced in the second and third goals of the syllabus (Kouluhallitus 1982):

"The goal is
- to familiarise the pupils in using mathematics for the analysis of their environment and solving problems,
- to develop the pupils' reasoning and evaluation skills and to promote the development of creativity".

The core syllabus, written for the basis of instructional planning and teaching materials, aimed at reducing the amount of changes required in instruction or material use.   That is why the contents remained distinctly more traditional than the objectives.   The new plan described the syllabus in such great detail that the generality of the curriculum was no longer criticised by the teachers.   The specification and simplification of the national curriculum had reached a certain extreme.

The development of the mathematics curriculum between 1970 and 1985 had the three phases presented above.   Figure 1 elucidates the timing of the different phases as well as how they have followed the international (American in particular) reforms of mathematics instruction. The development has many interesting sides to it.   One characteristic of interest can be seen in the shape, which indicates that new trends take a long time to reach Finland.   With regard to new mathematics the delay was about 10 years, whereas the emphasis on problem solving was acquired quite quickly.   Another significant feature has been the fact that in all phases, application and problem solving have been present either in the objectives or as aspects to be kept in mind in instruction.   Their weight in the curriculum has, perhaps, been continually increased, but they have not been thoroughly established in teaching practices.

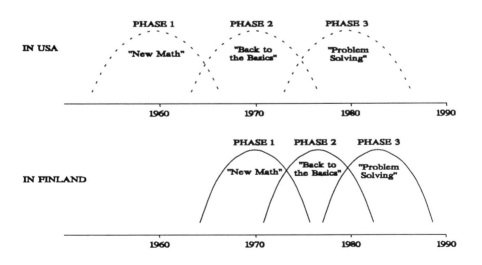

Figure 1 - The developmental phases of the comprehensive school mathematics curriculum in Finland related to the curricular trends in the Anglo-American world.

2.   IDEAS FOR DEVELOPMENT AFTER 1985

The new school laws which came into force in the fall of 1985 emphasise strongly the local right and responsibility for the development of the curriculum and school activities in general.   In practice this means, among other things, that the writing of the curriculum itself belongs to the municipality.   For this purpose the NBGE stipulated the principles for the comprehensvie school curriculum, where the mathematics syllabus again became more general and came from the 1970 curriculum.   The Cabinet's decision on the distribution of lessons per school year (1985) also had an effect on the contents of the mathematics syllabus, as the number of lessons during grade 9 was reduced.

From the point of view of mathematics instruction a greater change took place when sets (ability grouping) were removed at the upper level of the comprehensive school (grades 7–9).   Sets were replaced by inter–class differentiation, which is supported by the time–resource quota system.   The new system means that the schools are given a time-resource (depending on the size of the school), which the schools can use for the reduction of the number of pupils per teaching group for certain school subjects.   This reduction of the group sizes makes more individualised teaching possible.

Distribution of decisionmaking and the extensive changes in the structures of instruction have opened new and unpredictable situations for the school.   Why have we taken the road of decentralisation, and does it mean now that there are quite opposite trends taking place in some countries?   Does this have consequences for the position and role of the highest school administration which has lately arisen?   Shall we succeed in developing such solutions in mathematics instruction that make problem solving and applications instructional reality?

The idealisic aim of the NBGE was to turn the 1980s into the decade of problem solving.   Even though this had not progressed much by 1985, attempts have been made to proceed to a desirable end with the help of the development plan on mathematics instruction prepared by the NBGE.   One form of the development plan was to organise theme years.   The application of mathematical knowledge was the theme in the lower stage of the comprehensive school in the school year 1983–84 and was called Everyday Mathematics.   Correspondingly the following year was devoted to geometry.   The NBGE stressed the role of geometry in "using it in the structuring of the environment and in solving problems".   Problem solving was also a special theme in the plan, but only in the late 1980s.

The in–service education of mathematics teachers organised by the NBGE featured problem solving predominently in 1986–87.   Two national seminars on Problem Solving in Mathematics Instruction were arranged.   The idea of these seminars was to spread new information and impressions to teacher education as well as practical teaching.

Teachers have usually cited the lack of varied application tasks as the main reason for the paucity of applications in teaching. To improve this situation, the national ICMI Subcommittee decided, early in 1984, to start an application project in mathematics. The first concrete result of this project was a collection of application tasks taken from the textbooks of several countries. This material, for the comprehensive and upper secondary schools, was published in the autumn of 1988. The project is in progress and during 1989 teacher training colleges will, according to the plan, participate by requiring students to prepare seminar papers and essays on related topics.

Finally, I wish to mention a national blue—ribbon commission that is expected to contribute substantially to the development of mathematics instruction in Finland. The National Curriculum on Basic Education in Mathematics and Science, which was set up at the end of 1987, has already provided an evaluation of the current level of general knowledge of mathematics and science in Finland. Furthermore, the commission is expected to prepare a programme of activities for 1989—93 for the development of teaching at different levels of education. Undoubtedly, the commission will need to consider the role of applications, problem solving and modelling in the school mathematics of the 1990s and the beginning of the next century.

REFERENCES

Kouluhallits (1976). Matematiikka. Ehdotus perustavotteiksi ja perusoppiaineeksi peruskoulussa. (National Board of General Education: Mathematics. A proposal for core objectives and basic subject matter in the comprehensive school). Helsinki: Valtion painatuskeskus. In Finnish.

Kouluhallitus (1982). Peruskoulun matematiikan oppimäärä ja oppimääräsuunnitelma. (National Board of General Education: Mathematics syllabus and curriculum for the comprehensive school). Helsinki: Valtion painatuskeskus. In Finnish.

Leino, J. and Norlamo, P. (1980). Matematiikan kouluopetuksen sisältö tekniikan ja talouselämän näkökulmasta. (The content of school mathematics from the point of view of technology and economics). Matemaattisten aineiden opettajien liiton tutkimuksia N:0 1. In Finnish.

Peruskoulun opetussuunnitelmakomitean mietintö II (1970). Oppiaineiden opetussuunnitelmat. (The report of the Committee on the Comprehensive School Curriculum 1970. The Curricula of Subjects). Komiteanmietintö 1970:A5. Helsinki: Valtion painatuskeskus. In Finnish.

CHAPTER 30

The Teaching of Applications, Modelling and Problem Solving in Australia: 1984–1988

K. Stacey
Melbourne University, Australia
S. Groves
Victoria College, Australia

ABSTRACT

The intention of this brief review is to summarise developments in the teaching of applications, modelling and problem solving in Australian schools from 1984 to 1988. The aim is to point to major trends and to assist readers in finding further information, not to provide a comprehensive account of specific activities. As with almost all discussions of problem solving and applications, there is substantial difficulty in delineating and defining the area. In general a broad view has been adopted, although as far as possible *problem solving* has been reserved for activities where a significant degree of unfamiliarity characterises the task and *applications* is used when the interest lies in the process of applying appropriate mathematics and evaluating the usefulness of the results rather than in the teaching of mathematical technique alone. Similarly *modelling* is used to refer to teaching activities where the full modelling cycle of formulating, solving, interpreting and checking is of interest, rather than the mathematical techniques of the solving stage alone.

1. REFERENCE MATERIAL

Two articles summarise the work in problem solving in Australia prior to 1984. Mortlock (1988) takes a wide perspective, tracing early developments from the sixties, and Stacey & Groves (1984) give a comprehensive guide to work current in 1984. A third survey article (Groves & Stacey 1988) contains an annotated bibliography of 238 Australian articles on problem solving, applications and modelling published in the seventies and eighties. The bibliography includes articles within the tradition of academic research and, in an effort to trace the reasons and expectations for the current emphasis on problem solving, also includes a selection of articles more in the nature of exhortation to teachers. Along with information from curriculum committees in each state, the recent articles in the annotated bibliography formed the principal source of data for this review.

A further source of current information on problem solving, applications and modelling in Australia is the Problem Solving Interest Group of the Mathematics Education Research Group of Australasia. Contact addresses of this group and others mentioned below are given at the end of the article. A very useful set of references of activity in individual or small groups is provided by the proceedings of the annual conferences of the Mathematical Association of Victoria, the proceedings of the biennial conferences of the Australian Association of Mathematics Teachers and the mathematics teachers journals of each state.

2. THE OVERALL PICTURE

The years 1984 to 1988 have seen considerable change. Problem solving (including applications and modelling) has changed from a fashionable, fringe activity, which interested only a small number of teachers, to a formally recognised part of the curriculum aims of most schools and school systems. Problem solving, in one guise or another, has appeared as a strand in mathematics curricula across Australia, initially in primary and junior secondary courses, but more recently in the curriculum at year 12 level. In Queensland, the 1987 Guidelines document for the first 11 years of school describes learning activities in terms of concepts, processes and effects. Problem solving is seen as the global process, drawing on 13 general mathematical processes such as estimating, explaining and representing (Department of Education of Queensland 1987, p41). The Victorian 'Mathematics Framework' (Ministry of Education 1987) sees problem solving and applications instead as emerging issues (p15) along with others, such as group work, gender bias and technology. In Western Australia, process is assessed at every level, contributing 25% of the student's final mark in year 12 mathematics. The course Applying Mathematics for years 11 and 12 has six components, comprising application of essential skills, mathematical background and electives in which students study particular areas of application from a problem–solving point of view. In New South Wales, new senior mathematics courses have three categories of objectives, corresponding broadly to process and content. These are to give an understanding of important mathematical ideas, to develop understanding

of proof and to enhance skills required for further study.   Different
courses place different weights on these categories.   In Victoria, the new
year 11 and 12 mathematics courses propose three work requirements:
1.   skills practice and standard applications,
2.   problem solving and modelling,
3.   projects.

Students must spend at least 20% of their time on each work
requirement.   Included in the first work requirement are calculations such
as integrating standard functions, rearranging algebraic expressions and
applying mathematics to a range of common situations such as interest
rates, kinematics and navigation.   For the problem solving and modelling
work requirement, students will tackle questions taking from about 15
minutes to about eight hours.   Both pure mathematics investigations (such
as finding the total number of squares on a chessboard) and modelling
situations (such as developing a model which recommends the timing of
green and red lights at an intersection) are included.   Projects, usually
taking from about five to 20 hours, can be open-ended work of a
problem-solving or modelling nature, but they can also include descriptive
or historical accounts of mathematics and its uses.

During this period (which has unfortunately been one of severely
limited resources) an enquiry approach to teaching and the use of
realistic applications has been actively promoted by the state education
systems individually, and also collectively through the Mathematics
Curriculum of the States and Territories of Australia which supports the
improvement of teaching in mathematics in years K to 10.   It has
produced a bank of illustrative lesson ideas: practice-centred materials
which provide a window through which to glimpse how a particular
emphasis can be incorporated into classroom practice.   A problem-solving
approach to teaching underlies many of the lessons in the bank.   One
set of lessons collects together *Investigations and Problem Solving* and
one set is on *Mathematical Modelling*.   Each set of lessons is intended
to act as a vehicle by which teachers can explore a new territory such
as teaching modelling.   They do not provide a complete curriculum.

The period has also been one of substantial activity in many
individual schools, where teachers are anxious to respond to the call for
increased emphasis on problem solving and applications and to improve
the quality of the mathematical experiences offered to pupils.   The
teachers' journals and conference proceedings, referred to above, are the
best source of information on this.   One striking feature of the
grass-roots approach has been the spread of Mathematics Task Centres,
particularly for students in years 5 to 8.   The first Task Centre was
established by Neville de Mestre in Canberra in the seventies (de Mestre
1984).   The idea has now been extensively adapted to meet the needs of
different schools and groups of schools, but most variations seem to
retain three key features: tasks are concretely presented, they are stored
ready for use without further teacher preparation and students work on
the tasks in small groups.   A typical task is a variation of a

'man–wolf–cabbage' crossing the river' problem, written on a card and housed with appropriate concrete materials (such as a board, on which a river is drawn, and plastic figurines to represent the man, wolf and cabbage) housed in a clear plastic box on a shelf or trollies with perhaps 50 other such tasks.

A small but steady flow of research studies has continued throughout the period: many are annotated in Groves & Stacey (1988). Some of the research takes a cognitive science or artificial intelligence approach; other strands are challenged by the demands imposed by curriculum development and changing behaviour in average classrooms.

In teacher education, data collected informally by the Problem Solving Interest Group of MERGA indicated that problem solving is a formally recognised part of the compulsory mathematics methods courses for a large majority of primary teacher trainees. Secondary teacher trainees in four year concurrent courses are in a similar situation, and quite like meeting a problem–solving approach to both teaching and their own learning. However many trainee secondary teachers, who study education for a shorter time, meet a problem–solving approach to the teaching of ordinary content but have little exposure to problem solving in the sense of long investigations or modelling.

3.   CHALLENGES AHEAD

The picture painted above rightly reflects substantial activity and enthusiastic adoption by teachers, school systems and syllabuses of ideals of teaching which are espoused by the problem solving, applications and modelling movement. There is, of course, another side: our personal experience of visiting student teachers in fairly randomly chosen schools shows a rather different picture – time in most classrooms is still almost exclusively devoted to completing short practice exercises. From this perspective, problem solving and modelling seem to have been institutionalised without actually being adopted and 'acclaimed into impotence' as Bruner described the fate of new maths. While it is an important step ahead to change syllabus descriptions and official aims, it is quite another thing to change the learning and teaching that children experience in average schools.

Even so, the official changes already made pose very difficult challenges, especially in the areas of assessment, definitions of goals and descriptions of student progress and content/process balance. So many of the important characteristics of problem solving or modelling skills, for example, open a Pandora's box of difficult and fundamental questions, particularly when the assessment is for senior students about to leave school.   If it is important that students should plan their own investigations and follow their own paths, how can one student's work be compared with that of another student or of the same student when doing a different kind of investigation? The option that is being investigated in both Western Australia and Victoria is to outline a general set of criteria or guidelines for making any type of investigation, but many of the difficulties inherent in such schemes have not been overcome. To give one more example, another characteristic of problem–solving work valued by many is that students should undertake substantial pieces of work, perhaps lasting days or weeks. How, though, can long pieces of work be assessed fairly if students receive input from a variety of sources such as teacher, parents and friends? Assessing substantial pieces of work seems to fall inevitably on the teacher and the notion of teacher assessment is now broadly accepted. However, it is difficult to ensure that all teachers recognise and value the desired objectives in their students' work. Communicating to teachers the real aims of teaching for process is extremely difficult; it is still more difficult to communicate them to parents, potential employers and students.

The eagerness to embrace the new process objectives has lead to an artificial separation between content and process. In any task, mathematical content (knowledge of facts, concepts and skills) and the context of the problem interact with the process aspects of doing and applying mathematics. When this is not recognised, and attempts are made to assess process separately, it becomes impossible to define standards or progress. Thus, for example, the process of *formulating and testing hypotheses* (with which babies learn to speak their native language) might also be listed as an objective of year 12 mathematics. Finally, there is a danger that we may begin to assess the unteachable – until we know how our teaching helps students to improve, is it fair to assess them?

In a summary of the work of the Problem Solving Theme Group at ICME–5, Stacey & Groves (1984a) make 22 recommendations about work needed to advance problem solving in schools. The recommendations deal with the need for good teaching materials, the need to investigate and promote appropriate teaching styles, teacher education for problem solving, definition of values and goals, the need of syllabus statements to recognise process objectives and the conditions under which they may be achieved, the need for research into questions arising from practice, and the need for system–wide support. Since that time, Australia has seen a steady stream of good teaching materials (a list is available from the Problem Solving Interest Group) and there are good prospects of filling the large and important gaps which are left. The teacher development work of the MCTP has made good, organised progress in supporting teachers who wish to extend their teaching style; this is clearly an on–going task and it will be some time before it bears fruit in a significant number of classrooms. Syllabus statements have now recognised process as well as content and teachers who seek it out can usually find adequate, and very often enthusiastic, support from the system for their efforts in this area. Currently the most urgent questions are those arising from assessment, particularly for senior mathematics courses. What is it that we really value and how can it best be achieved within the constraints of ordinary school life?

REFERENCES

de Mestre, N. (1984). The Australian Capital Territory Mathematics Centre. In Costello, P., Ferguson, S., Slinn, K., Stephens, M., Trembath, D. and Williams, D. (eds) *Facets of Australian Mathematics Education*. Blackburn, Victoria: Australian Association of Mathematics Teachers, 221-224.

Department of Education of Queensland (1987). *Years 1 to 10 Mathematics: Teaching, Curriculum and Assessment Guidelines*. Brisbane, Queensland.

Groves, S. and Stacey, K. (1988). Problem Solving: An Annotated Bibliography. In Blane, D. and Leder, G. (eds) *Mathematics Education in Australia: A Selection of Recent Research*. Melbourne, Victoria: Mathematics Education Research Group of Australasia, 66-119.

Ministry of Education of Victoria (1987). *The Mathematics Framework P - 10: A forward look*. Melbourne.

Mortlock, R. (1988). Problem Solving - The State of the Art in Australia. In Burkhardt, H., Groves, S., Schoenfeld, A. and Stacey, K. (eds) *Problem Solving - A World View*. Nottingham, England: Shell Centre for Mathematical Education, 144-150.

Stacey, K. and Groves, S. (1984). Problem Solving: People and Projects in Australia. In Costello, P., Ferguson, S., Slinn, K., Stephens, M., Trembath, D. and Williams, D. (eds) *Facets of Australian Mathematics Education*. Blackburn, Victoria: Australian Association of Mathematics Teachers, 205-209.

Stacey, K. and Groves, S. (1984a). Problem Solving: The Way Forward after ICME 5. In Maurer, A. (ed) *Conflicts in Mathematics Education*. Melbourne: Mathematical Association of Victoria, 397-405.

CONTACT ADDRESSES

Mr Noel Thomas
Secretary, Problem Solving Interest Group,
Mathematics Education Research Group of Australasia,
Mitchell College of Advanced Education,
Bathurst, NSW, Australia.

AAMT National Services Agent,
Australian Association of Mathematics Teachers,
18 Bluehills Road,
O'Halloran Hill 5158
South Australia, Australia.

Mathematical Association of Victoria,
191 Royal Parade,
Parkville 3052, Victoria, Australia.

Mathematics Curriculum and Teaching Project,
Curriculum Development Centre,
PO Box 34,
Woden 2606,
ACT, Australia.

Index

Mathematics and its Applications

Series Editor: G. M. BELL, Professor of Mathematics, King's College London (KQC), University of London

Gardiner, C.F.	**Algebraic Structures**
Gasson, P.C.	**Geometry of Spatial Forms**
Goodbody, A.M.	**Cartesian Tensors**
Goult, R.J.	**Applied Linear Algebra**
Graham, A.	**Kronecker Products and Matrix Calculus: with Applications**
Graham, A.	**Matrix Theory and Applications for Engineers and Mathematicians**
Graham, A.	**Nonnegative Matrices and Applicable Topics in Linear Algebra**
Griffel, D.H.	**Applied Functional Analysis**
Griffel, D.H.	**Linear Algebra and its Applications: Vol. 1, A First Course; Vol. 2, More Advanced**
Guest, P. B.	**The Laplace Transform and Applications**
Hanyga, A.	**Mathematical Theory of Non-linear Elasticity**
Harris, D.J.	**Mathematics for Business, Management and Economics**
Hart, D. & Croft, A.	**Modelling with Projectiles**
Hoskins, R.F.	**Generalised Functions**
Hoskins, R.F.	**Standard and Non-standard Analysis**
Hunter, S.C.	**Mechanics of Continuous Media, 2nd (Revised) Edition**
Huntley, I. & Johnson, R.M.	**Linear and Nonlinear Differential Equations**
Irons, B. M. & Shrive, N. G.	**Numerical Methods in Engineering and Applied Science**
Ivanov, L. L.	**Algebraic Recursion Theory**
Johnson, R.M.	**Theory and Applications of Linear Differential and Difference Equations**
Johnson, R.M.	**Calculus: Theory and Applications in Technology and the Physical and Life Sciences**
Jones, R.H. & Steele, N.C.	**Mathematics in Communication Theory**
Jordan, D.	**Geometric Topology**
Kelly, J.C.	**Abstract Algebra**
Kim, K.H. & Roush, F.W.	**Applied Abstract Algebra**
Kim, K.H. & Roush, F.W.	**Team Theory**
Kosinski, W.	**Field Singularities and Wave Analysis in Continuum Mechanics**
Krishnamurthy, V.	**Combinatorics: Theory and Applications**
Lindfield, G. & Penny, J.E.T.	**Microcomputers in Numerical Analysis**
Livesley, K.	**Mathematical Methods for Engineers**
Lord, E.A. & Wilson, C.B.	**The Mathematical Description of Shape and Form**
Malik, M., Riznichenko, G.Y. & Rubin, A.B.	**Biological Electron Transport Processes and their Computer Simulation**
Massey, B.S.	**Measures in Science and Engineering**
Meek, B.L. & Fairthorne, S.	**Using Computers**
Menell, A. & Bazin, M.	**Mathematics for the Biosciences**
Mikolas, M.	**Real Functions and Orthogonal Series**
Moore, R.	**Computational Functional Analysis**
Moshier, S.L.B.	**Methods and Programs for Mathematical Functions**
Murphy, J.A., Ridout, D. & McShane, B.	**Numerical Analysis, Algorithms and Computation**
Nonweiler, T.R.F.	**Computational Mathematics: An Introduction to Numerical Approximation**
Norcliffe, A. & Slater, G.	**Mathematics of Software Construction**
Ogden, R.W.	**Non-linear Elastic Deformations**
Oldknow, A.	**Microcomputers in Geometry**
Oldknow, A. & Smith, D.	**Learning Mathematics with Micros**
O'Neill, M.E. & Chorlton, F.	**Ideal and Incompressible Fluid Dynamics**
O'Neill, M.E. & Chorlton, F.	**Viscous and Compressible Fluid Dynamics**
Page, S. G.	**Mathematics: A Second Start**
Prior, D. & Moscardini, A.O.	**Model Formulation Analysis**
Rankin, R.A.	**Modular Forms**
Scorer, R.S.	**Environmental Aerodynamics**
Shivamoggi, B.K.	**Stability of Parallel Gas Flows**
Smith, D.K.	**Network Optimisation Practice: A Computational Guide**
Srivastava, H.M. & Manocha, L.	**A Treatise on Generating Functions**
Stirling, D.S.G.	**Mathematical Analysis**
Sweet, M.V.	**Algebra, Geometry and Trigonometry in Science, Engineering and Mathematics**
Temperley, H.N.V.	**Graph Theory and Applications**
Temperley, H.N.V.	**Liquids and Their Properties**
Thom, R.	**Mathematical Models of Morphogenesis**
Toth, G.	**Harmonic and Minimal Maps and Applications in Geometry and Physics**
Townend, M. S.	**Mathematics in Sport**
Townend, M.S. & Pountney, D.C.	**Computer-aided Engineering Mathematics**
Trinajstic, N.	**Mathematical and Computational Concepts in Chemistry**
Twizell, E.H.	**Computational Methods for Partial Differential Equations**
Twizell, E.H.	**Numerical Methods, with Applications in the Biomedical Sciences**
Vince, A. and Morris, C.	**Mathematics for Computing and Information Technology**
Walton, K., Marshall, J., Gorecki, H. & Korytowski, A.	**Control Theory for Time Delay Systems**
Warren, M.D.	**Flow Modelling in Industrial Processes**
Wheeler, R.F.	**Rethinking Mathematical Concepts**
Willmore, T.J.	**Total Curvature in Riemannian Geometry**
Willmore, T.J. & Hitchin, N.	**Global Riemannian Geometry**

Statistics, Operational Research and Computational Mathematics

Editor: B. W. CONOLLY, Emeritus Professor of Mathematics (Operational Research), Queen Mary College, University of London